Climate Change in the Indo-Pacific

This book explores multifarious policy dimensions relevant to climate change and its associated vulnerabilities in the Indo-Pacific region.

The chapters in the book focus on:

- climate resilience in terms of policy implementation at the local, national, and regional levels
- understanding climate change through the lens of non-traditional security
- assessing geo-economic and geopolitical impacts and adaptation strategies
- examining issues of climate refugees and migration

The book will be of great interest to scholars and researchers of politics, public policy, environment and sustainability studies, climate action, and Indo-Pacific studies.

Faisal Ahmed is Professor in the International Business Area teaching trade and geopolitics at FORE School of Management, New Delhi, India.

Mohd. Faheem is Assistant Professor in Indian Studies at the Pridi Banomyong International College, Thammasat University, Thailand.

Indo-Pacific in Context

This series brings together topical research on contemporary and long-standing issues encompassing the Asia-Pacific region.

With countries steeped in history, communities diverse in cultures, developing economies and emerging markets, Asia-Pacific has emerged as the key stakeholder in a world order in flux. The region has solidified its presence in the global political discourse through multilateral initiatives, defence agreements, and strategic partnerships. It has emerged as a zone of contestations, conflict, and cooperation.

The works published in this series showcase interdisciplinary research in the arts, the humanities, and the social sciences, including a range of subject areas such as politics and international relations, international economy, sociology and social anthropology, women, gender and sexuality studies, history, geo-politics, military studies, area studies, cultural studies, environment and sustainability, development studies, migration studies, urban development, digital humanities, and science and technology studies.

Works in the series are published simultaneously in UK/US and South Asia editions, as well as in e-book format. We welcome a range of books aimed at furthering scholarship and understanding of the Asia-Pacific region. Authors and researchers interested in contributing to this series may get in touch with rioeditorial@tandfindia.com

Handbook of Indo-Pacific Studies
Edited by Barbara Kratiuk, Jeroen J. J. Van den Bosch, Aleksandra Jaskólska and Yoichiro Sato

Japan and its Partners in the Indo-Pacific
Engagements and Alignment
Edited by Srabani Roy Choudhury

The New World Politics of the Indo-Pacific
Perceptions, Policies and Interests
Edited by Josukutty C A and Joyce Sabina Lobo

Climate Change in the Indo-Pacific
Developmental and Geopolitical Dimensions
Edited by Faisal Ahmed and Mohd. Faheem

For more information about this series, please visit: www.routledge.com/Indo-Pacific-in-Context/book-series/IPC

Climate Change in the Indo-Pacific

Developmental and Geopolitical Dimensions

Edited by Faisal Ahmed and Mohd. Faheem

Routledge
Taylor & Francis Group
LONDON AND NEW YORK

First published 2025
by Routledge
4 Park Square, Milton Park, Abingdon, Oxon OX14 4RN

and by Routledge
605 Third Avenue, New York, NY 10158

Routledge is an imprint of the Taylor & Francis Group, an informa business

British Library Cataloguing-in-Publication Data
A catalogue record for this book is available from the British Library

ISBN: 978-1-032-52188-6 (hbk)
ISBN: 978-1-032-98811-5 (pbk)
ISBN: 978-1-003-58686-9 (ebk)

DOI: 10.4324/9781003586869

Typeset in Sabon
by Deanta Global Publishing Services, Chennai, India

Contents

Contributors

Faisal Ahmed is Professor in the International Business Area teaching trade and geopolitics at FORE School of Management, New Delhi, India. In addition to a Doctorate and Master's from India, he holds an executive certification in geopolitical analysis from Geneva. His research and consulting interests include Indo-Pacific region, China's trade and geopolitics, trade agreements, and climate change. He had been Consultant to the United Nations and other organisations. He has also worked on projects supported by the Ministry of Commerce and Industry, Government of India, and others. He has published papers in refereed journals and frequently publishes in popular media. He is the author of the textbook *Business Environment: Indian and Global Perspective* (2014). He has published three edited books and three monographs on grassroots reach of India's foreign trade policy, India's experiences in PTAs, and EEZ and the UN Convention on the Law of the Sea. His articles have appeared in *The Times of India*, *The Korea Times*, *The Straits Times*, *South China Morning Post*, *The Hindu Business Line*, and *The Economic Times*, among others. He is frequently interviewed as a geopolitical expert by national and international media.

Anju has worked as Assistant Professor (Guest Faculty), Aryabhatta College, Janki Devi Memorial College, Shri Guru Gobind Singh College of Commerce, NCWEB University of Delhi from September 2019 to April 2021. She has completed her Graduation from Lady Shri Ram College for Women and Post-graduation in History from Kirori Mal College, University of Delhi. She has done M.Phil and Ph.D degrees from Centre for South Asian Studies, School of International Studies, Jawaharlal Nehru University, New Delhi. She has published many research articles and papers in national journals related to South Asian History. She has also presented many research papers in national and international conferences/ seminars. Her latest research work is "Social Reforms on Modern India", which has been published in Alagappa University. Her research interest areas are South Asian History, women empowerment, status of Dalits, and migration issues in South Asia.

Aparajita is Assistant Professor in the Department of Political Science, B. D. College, Patliputra University, Patna (Bihar). She has done her M.Phil and Ph.D from the Centre for South Asian Studies, School of International Studies, Jawaharlal Nehru University, New Delhi, under the supervision of Prof. Amita Batra. The topic of her thesis is "South-Asia and Anti-Dumping Dispute Settlement at the World Trade Organisation". She has presented papers at several international and national conferences and has few publications in reputed journals. Her areas of interests are South Asia, Foreign Policy, SAARC, World Trade Organisation, Regional Integration, Dispute Settlement Mechanism, and International Relations Theories.

Maya Dania is Assistant Professor of Sociology at the School of Social Innovation, Mae Fah Luang University, Thailand. She holds a Bachelor of Philosophy from Gadjah Mada University in Indonesia and an M.A. in Southeast Asia Studies from Chulalongkorn University in Thailand. She is currently working on her Ph.D in Social Science at Chiang Mai University, focusing on "Anthropocene, Eco-Feminism, and Multispecies". She has contributed a number of research articles in the SCI Journal on gender violence and gender disaster issues. She was a fellow at the National University of Singapore in 2016 and Malaya University in 2016.

Mohd. Faheem is Assistant Professor in Indian Studies at the Pridi Banomyong International College, Thammasat University, Thailand. He holds Master's and Ph.D degrees in Geography from Aligarh Muslim University, India. Dr Faheem's research areas include India's relations with ASEAN, international migration, urban geography, economic geography, and political geography. He used to work as Visiting Research Fellow at the Institute of Asian Studies, Chulalongkorn University. He has also taught as Guest Lecturer at Srinakhrinwirot University and Bangkok Thonburi University. He has published several articles in international referred journals and presented papers in conferences and seminars. Dr Faheem has served as resource person in many conferences. He has published book chapters and edited two books. He has given his opinions and interviews on India in Thai newspapers and magazines. He is a Member of the advisory committee of Sir Syed Global Excellence Awards. He is also a member of Thailand and Malaysia Think Tank Scholars Network and Scholars Network for Peace. Dr Faheem also serves as member of the editorial and advisory board in peer-reviewed journals.

Reni Juwitasari is Research Associate at Asian Research Center for International Development (ARCID), School of Social Innovation, Mae Fah Luang University. She holds a Master's Degree in Educational Administration from Mahasarakham University, Thailand. She has published several articles on disaster and women's leadership and girls in education. She is currently conducting research on sustainability and sustainable development on Japan's Official Development Assistance in

Mekong Sub-Region country. She is also a Member of as part of UNESCO Chair Kobe University on Gender and Vulnerability in Disaster Risk Reduction.

Edmund Komboli is Senior Foreign Services Officer in the Department of Foreign Affairs, Government of Papua New Guinea, Port Moresby. He hold a degree in Political Science and Public Policy Management from the University of Papua New Guinea and a Master's degree in International Relations from the International University of Japan.

Alexandre Lambert is Academic Director of the Geneva Institute of Geopolitical Studies (GIGS), Switzerland. He is also Director of a SIT Study Abroad University Program on Global Health and Development Policy in Switzerland and teaches international relations, with focus on international security at the Geneva School of Diplomacy and International Relations (GSD) and International University Geneva (IUG). Formerly, he was a Civil Servant at the Swiss Federal Department of Defence within its security policy division and became a founding fellow of the Geneva Centre for Security Sector Governance (DCAF). He holds a Ph.D from the Graduate Institute of International and Development Studies in Geneva, where he was also Lead Researcher on the OSCE, to which he has been providing independent policy advise for almost two decades and with special emphasis on the Code of Conduct on Politico-Military Aspects of Security. He has mainly published in the fields of international security, civil–military relations, and geopolitics.

Aslam Perwaiz is Deputy Executive Director (DED) of Asian Disaster Preparedness Center (ADPC). Holding a Master's degree in Statistics with acquired expertise on disaster risk management. He is currently leading program development, funds acquisition, and program management of ADPC thematic areas including climate resilience, risk governance, urban resilience, health risk management, preparedness for response and recovery, and geospatial information. He has been involved in various global and regional initiatives on disaster resilience, business continuity planning, and digital resilience. Among other notable initiatives, he mentors the ADPC's privates sector engagement "iPrepare Business" facility and post-disaster needs assessment initiative "Ready4Recovery". Prior to joining ADPC, he worked with the United Nations Development Program (UNDP), World Wildlife Fund (WWF), and the Institute of Economic Growth (IEG) in India.

Rajendra Prasad Patel is Assistant Professor Political Science at Kunwar Singh PG College, Ballia. He has worked as Assistant Professor (Guest Faculty), Aryabhatta College and Hansraj College, NCWEB, University of Delhi from 10th October 2019 to 13th March 2020. He holds a BA Degree from Rajdhani College and MA from Kirori Mal College, University of Delhi. He has M.Phil and PhD Degrees from Centre for South Asian

Studies, School of International Studies, Jawaharlal Nehru University and he has also obtained Junior Research Fellowship. He has published many research articles in national journals related to India–Bangladesh trade and connectivity. Her latest article is "Extremist Upsurge in Bangladesh Politics", which has been published in *Journal of North East India Studies*. He has also presented many research papers in national and international conferences/seminars. His areas of interest are regional economic cooperation, intra-regional transport connectivity and climate change in South Asia, and India–Bangladesh economic cooperation.

Pushkar Pushp is Assistant Professor in the International Business Area at FORE School of Management, New Delhi. His area of research expertise encompasses global value chains and sustainability. His paper has been published in journal indexed in ABDC-A, SSCI and Scopus. Furthermore, his articles have appeared in magazines such as *The Geopolitics* and *Diplomatist*. Prior to his academic pursuits, he has also worked with BOSE Corporation Pvt Ltd.

Swasti Rao is Associate Fellow, Europe and Eurasia Center, Manohar Parrikar Institute for Defence Studies and Analyses, New Delhi, India. She has a Master's in Politics with specialisation in International Relations and an M.Phil. in East Asian Studies from Jawaharlal Nehru University, New Delhi. She has a PhD in Advanced International Politics from Tsukuba University, Japan, where she lived for five years on the prestigious Japanese Education Ministry Fellowship (MEXT) that she received in 2011. Her PhD thesis is titled, "Civil Society and Democracy Theory: A Comparative Study of Japan and India". She is also a recipient of the 'Okita Memorial Scholarship' in 2010 from ICCR and has published original research under those auspices. She has also qualified the UGC NET in International and Area Studies in 2007. Her academic and research interests focus on Participatory Democracy, Civil Society, Research Methodology and International Relations Theory. An expert on East Asian politics, she has been working on the traditional as well as the non-traditional threats in contemporary geopolitics particularly in context of China's rise. She has also served as an Assistant Professor at the Department of Security and Strategy Studies, Aligarh Muslim University, Aligarh, India, where she offered courses at Undergraduate, Master's and PhD levels.

Pittaya Suvakunta is Associate Professor at College of Interdisciplinary Studies, Thammasat University, Thailand. He holds Ph.D (Integrated Science) from Thammasat University, which was funded by The Royal Golden Jubilee PH.D. Program, the Thailand Research Fund. He served as Deputy Dean of Academic and Research (2013–2016) at College of Interdisciplinary Studies, Thammasat University. He has been a Visiting Professor at Vietnam National University, University of Humanity and

Social Science. His research interests include Area Studies about Chinese Studies, and Southeast Asia Studies.

Shweta Sinha is Assistant Professor of Environmental Technology at Pridi Banomyong International College, Thammasat University, Thailand. She has more than 15 years of international experience in projects, research, consultancies, and capacity building in the area of climate change, disaster risk management, digital agriculture, community resilience, and gender and sustainable development. Her Ph.D. is in the field of agriculture risk management and financing from the Faculty of Remote Sensing and GIS at Asian Institute of Technology, Thailand. Prior to Ph.D., she has worked for Siemens IT Solutions and Services, Thailand, now ATOS, in the field of capacity building, service management and support of SAP and for TCIL Bellsouth Ltd, India, in the field of telecom software development. She also holds a MBA degree in International Business and an Engineering degree in Electronics.

Introduction

Faisal Ahmed and Mohd. Faheem

The Indo-Pacific region is particularly prone to natural disasters, a vulnerability compounded by the escalating impacts of climate change. The rise in more frequent and intense extreme weather events, along with the challenge of sea-level rise, coastal inundation, coral reef damage, increased incidence of natural disasters, water insecurity, and health-related epidemics like water and vector-borne diseases, collectively contribute to an acute crisis in the region. For instance, 2021 witnessed the occurrence of 174 natural disasters in the Asian region alone, affecting approximately 66.8 million individuals, with over 12 million facing displacement in East Asia and the Pacific region. Notably, from 2017 to 2021, an estimated 36,000 lives were lost due to the impact of natural disasters.[1]

A major part of South Asia and Southeast Asia faces heightened susceptibility to El Niño and La Niña events, with both phenomena exhibiting escalating severity. This severity is manifested by augmented precipitation during wet years, increased cyclone risk, and aggravated aridity during dry periods. Presently, the region finds itself in a precarious state, exacerbated by the intensification of the current La Niña, which has resulted in unprecedented flooding in locales still recovering from prior extreme conditions, including intense heat, drought, and food security shocks (Glasser et al., 2022).

Similarly, the Pacific Island states are witnessing disruptions due to climate change. Small Island nations such as Fiji, Kiribati, and Tuvalu find themselves particularly vulnerable due to the ongoing repercussions of climate change (Ahmed & Mishra, 2020). The rise in sea levels and the occurrence of storm surges present catastrophic challenges for low-lying areas in the Pacific islands, including Kiribati and Tuvalu. A study conducted by Merone and Tait (2018) further forecasts that both Kiribati and Tuvalu are anticipated to become uninhabitable by the year 2050, underscoring the severe consequences of climate-induced environmental changes in the Indo-Pacific region. Due to the sudden spike in natural disasters and repercussions pertaining to climate change, the concept of 'loss and damage' was rooted in the United Nations Paris Agreement.

Moreover, climate change has the potential to reshape security dynamics within the Indo-Pacific region, giving rise to increased instances of piracy and transnational criminal activities. These phenomena are attributed to

DOI: 10.4324/9781003586869-1

climate-induced factors such as displacement, poverty, and overfishing stemming from ocean warming and acidification. According to projections by the United Nations, the Indo-Pacific region may witness a substantial influx of climate refugees, with estimates reaching as high as 89 million by the year 2050. Notably, the majority of these climate refugees are anticipated to originate from Southeast Asia and the Pacific.[2]

Several nations within the Indo-Pacific region exhibit weak governance structures marked by corruption and diminished levels of legitimacy in governmental institutions (The World Bank, 2019; Tangney et al., 2021). Alongside, persistent safety and human security issues arise in countries like Malaysia, Indonesia in Southeast Asia, and Pacific Island states such as Tonga and Vanuatu. These concerns pose a potential threat to regional security through implications such as forced migration and resource conflicts. Thus, heightened pressures resulting from extreme events may undermine the capacity of governments in the region to efficiently coordinate resources and personnel (Tangney et al., 2021).

Over the past two decades, regional institutions and organisations such as ASEAN, the Pacific Islands Forum, and multilateral development banks have elevated their endeavours to address the diverse economic, political, social, security, and humanitarian challenges stemming from the warming climate. Notably, in the Pacific, these entities have played a pivotal role in fortifying climate resilience and facilitating the relocation of island communities displaced by rising sea levels and more potent cyclones. However, their initiatives are increasingly overwhelmed by the magnitude of climate hazards and their cascading repercussions. Efforts to advocate for regional strategies in addressing these crises face considerable impediments, with certain states opting to fortify barriers and adopt insular approaches instead (Glasser et al., 2022).

The impacts of climate change have reached an unprecedented level of severity and frequency, leading to the incorporation of the notion of 'loss and damage' into the Paris Agreement of the United Nations. The conclusion of the United Nations Climate Change Conference (COP28) marked a significant milestone, signalling the 'beginning of the end' of the fossil fuel era.[3] This accord establishes the foundation for a rapid, just, and equitable transition, characterised by substantial reductions in emissions and amplified financial commitments. Delegates representing nearly 200 Parties convened in Dubai and reached a consensus on the initiation of the world's inaugural 'global stocktake', aimed at intensifying climate action prior to the decade's conclusion, with the overarching objective of maintaining the attainability of the global temperature threshold of 1.5°C. Parties further endorsed targets for the Global Goal on Adaptation (GGA) and its accompanying framework, outlining the resilience measures to combat the impacts of a shifting climate and evaluating nations' mitigation endeavours.

The Indo-Pacific region assumes significance as a geostrategic location, along with geopolitical tensions and competition among major powers. There exists a pressing need for the region to proactively address climate security concerns. China, the world's leading emitter of greenhouse gas emissions, has committed to incorporating nitrogen oxide in its 2035 national climate plan. Additionally, China has outlined specific measures to curb methane emissions. Noteworthy is China's existing workforce of approximately 54 million in the 'green jobs' sector, with over 4 million engaged in renewable energy. China has also declared a cessation of coal-fired power plant construction overseas and an augmented commitment to supporting other nations in developing green and low-carbon energy sources.

The US too has also set ambitious targets to tackle climate change at home and internationally. For example, the ambitious target to achieve carbon neutrality by reducing greenhouse gas emissions by 50–52 per cent by 2030 and the goal for the share of electric vehicle sales to be 50 per cent by 2030 are some of the measures adopted by the US. Moreover, the Inflation Reduction Act (IRA) is expected to cut down Greenhouse Gas (GHG) emissions by 1 billion tons by 2030. In fact, to tackle the GHG emissions coming from the maritime industry, the US along with Norway, has launched a green shipping challenge to confront climate change.

India too has declared its commitment to achieving net-zero emissions by the year 2070. During the 26th session of COP 26 in November 2021, India submitted its long-term low-carbon development strategy to the UNFCCC. This strategy reaffirms the objective of attaining net-zero emissions by 2070.

Importantly, the Asia-Pacific Adaptation Network (APAN) serves as an important platform for providing governmental officials and key stakeholders with essential knowledge for formulating strategies related to climate change adaptation. It facilitates the acquisition of resources, such as finance and technologies, along with enhancing the capacity of individuals involved in integrating climate change adaptation into the framework of national development policies. Operating within the ambit of the Global Adaptation Network—an extensive initiative by the United Nations Environment Programme (UNEP)—APAN adopts a collaborative approach as a 'network of networks'. The primary objective is to effectively manage and disseminate knowledge pertaining to climate change adaptation, aligning with the practical requirements of on-the-ground climate change practitioners. The network actively contributes to the establishment of climate-resilient and sustainable human systems, ecosystems, and economies within countries across the region. APAN endeavours to safeguard individuals, livelihoods, and ecosystems. While numerous nations and communities have initiated measures to fortify their societies and economies against climate-related challenges, a greater level of concerted action and ambition is imperative to mitigate present and future risks.

This edited book delves deep into understanding the climate change-related issues, challenges, and policy imperatives in the Indo-Pacific region. The same has been discussed in Chapter 1 of the book, which highlights that synergy between policies, programmes, and institutions; strengthened coordination mechanisms and partnerships; and a breakdown of traditional silos at the national and sub-national level will ensure the effectiveness, efficiency, achievability, and sustainability of the global frameworks envisaged for climate change. As outlined earlier, the Indo-Pacific remains an epicentre of geopolitical tensions and climate-related disasters. Chapter 2 critically analyses the impact of geopolitics on climate change. In fact, this chapter dives deep into assessing the geostrategic policy agendas and power rivalries and explains the different narratives by superpowers to tackle climate change. This chapter also highlights the environmental viability of the Belt and Road Initiative (BRI). Covering the geopolitical and geo-economic impact of climate change and adaptation strategies by region to tackle climate change, Chapter 3 aims to study the strategies of countries in the Asia-Pacific region to tackle climate change. It highlights that climate change is expected to affect many sectors and present risks to many Asia-Pacific Economic Cooperation (APEC) economies in Asia. Chapter 4 of this book highlights some of the best practices that may be used by Pacific Communities to improve their resilience to the impact of climate change, sea-level rise, and extreme weather events with reference to coastal adaptation measures. The discussion on the securitisation of climate change is presented in Chapter 5. This chapter highlights that climate change is one of the most tangible security challenges of our times. The effort to securitise the issue of climate change has sped up since the last decade owing to its increasing existential threat. The securitisation of climate change and strengthening an international regime complex to tackle its challenges has to be done in both theory and practice.

Climate-related incidents are quite common in Southeast Asia as well. Chapter 6 discussed that Southeast Asia is located in one of the world's most disaster-prone regions, causing almost all types of natural hazards, predominantly earthquakes. In fact, between 1970 and 2009, earthquakes affected 114,000 casualties, which led to the attention of all government levels in an attempt at disaster risk reduction, particularly the local government. The local government in the province needs to promote innovative and collaborative approaches to enhance effectiveness in disaster risk activities. A set of disaster management policies has been established at the provincial level through a review of its emergency preparedness and disaster response operations, reflected in the National Constitution or Act on Disaster Management, which is derived from the international platforms United Nations Office for Disaster Risk Reduction (UNISDR) and commitment to the Hyogo Framework of Action (HFA). However, while some policies were well implemented, others still lagged due to a predicament in collaboration that always transpires from different outlooks and readiness. Using Thailand, which represents mainland Southeast Asia, and Indonesia, which represents Southeast Asia's islands, this research aimed to investigate how the local governments in Thailand and Indonesia performed Disaster Risk Governance (DRG) with a case study from the 2014 Chiang Rai earthquake and the 2006 Yogyakarta earthquake.

In the Indian Ocean islands also, climate change poses serious problems and risks to global people, ecosystems, and the economy. Efforts for reducing greenhouse gases need to be complemented by multiple stakeholders and their responses to build resilience. Chapter 7 discusses adaptation and mitigation strategies to tackle climate change in Indian Ocean islands.

Moreover, climate refugees are emerging as a major humanitarian challenge caused by climate change. Sea-level rise, coastal inundation, natural disasters, water insecurity, and health-related epidemics in the form of water and vector-borne diseases, among others, have contributed immensely to this critical humanitarian crisis. Climate refugees are also lacking protection within the frameworks of international law and, more so within the domestic political systems of refuge-granting countries. Chapter 8 thus delves deep into assessing the factors that are causing the climate refugee crisis in the Indo-Pacific region, which includes countries sharing borders with the Indian Ocean as well as the Pacific Ocean. It also assesses the provisions of international legal frameworks, relevant inter-governmental conventions (e.g., the UNFCCC), and outlines the legal and technical impediments. The increasing severity of the climate refugee crisis will, in fact, affect the social structure and deprive the refugees of their ability to sustain livelihoods. This chapter also presents pragmatic policy and social interventions needed to address the climate refugee crisis.

Chapter 9 discusses climate-induced forced migration in South Asia. This region is one of the most affected regions of the world. The region's different geography combines with regional circulation trends to create a diverse climate. The region has both hot and cold extreme weather that are challenging for human well-being and climate change. The chapter highlights how vulnerable groups are currently experiencing the effects of climate on food, water, and other ecosystem services necessary for human well-being, livelihood, settlement, and survival. Women, children, older people, people with disabilities, and indigenous people are disproportionately influenced.

Chapter 10 highlights that Bangladesh has been recognised internationally as a country most vulnerable to climate change. According to the UN Mortality Risk Index, Bangladesh is mostly vulnerable due to cyclones and floods. The rise in temperature, erratic rainfall, precipitation due to monsoons, moisture stress during dry periods, rise in sea level, salinity increase, and increase in the density of cyclones are some of the adverse impacts of climate change that hinder Bangladesh's development. It negatively affects the food and water security, people's health, economy, and the environment of Bangladesh. There is a need for Bangladesh to adopt measures to deal with such vulnerabilities. Bangladesh has adopted several policies both at the national and international level to deal with the issue of environmental vulnerabilities in general and climate change in particular.

The book would thus serve as a reference guide for policymakers, think-tanks, and academics in developing an improved understanding of the climate change challenges and vulnerabilities in the Indo-Pacific region.

Notes

1 https://www.eastasiaforum.org/2022/12/26/securitising-climate-policy-will-keep-the-indo-pacific-afloat/
2 https://reliefweb.int/report/indonesia/shore-thing-climate-change-and-maritime-security-intrinsically-linked#:~:text=The%20United%20Nations%20estimates%20that,due%20to%20rising%20sea%20levels.
3 ttps://www.lowyinstitute.org/the-interpreter/understanding-loss-damage-climate-change-across-indo-pacific

References

Ahmed, F., & Mishra, V. (2020). Estimating relative immediacy of water-related challenges in Small Island Developing States (SIDS) of the Pacific Ocean using AHP modeling. *Modeling Earth Systems and Environment*, 6(1), 201–214.

Glasser, R., Johnstone, C., & Kapetas, A. (Eds.). (2022). *The geopolitics of climate and security in the Indo-Pacific*. Australian Strategic Policy Institute.

Merone, L., & Tait, P. (2018). Preventing disaster in the Pacific Islands: The battle against climate disruption. *Australian and New Zealand Journal of Public Health*, *42*(5), 419–420.

Tangney, P., Nettle, C., Clarke, B., Newman, J., & Star, C. (2021). Climate security in the Indo-Pacific: A systematic review of governance challenges for enhancing regional climate resilience. *Climatic Change*, *167*(3–4), 40.

The World Bank. (2019). *Implementation completion and results report for building climate/disaster resilience along the Dili-Ainaro and linked road corridors in Timor-Leste (P144818)*. World Bank.

1 Policies and Practices on Coherence between Disaster and Climate Resilience in Asia and the Pacific

Aslam Perwaiz[1] *and Shweta Sinha*[2]

Disaster Impacts in Asia and the Pacific

The countries in Asia and the Pacific have immense potential for future growth and have already demonstrated impressive economic development during the past decade. Asia has been estimated to become the home of half of the world's middle class, and the countries in the region continue to be among the fastest-growing economies globally. When combined, they may be larger than the rest of the world by 2020 (Romel & Reed, 2019). Many of the countries, including Lao PDR, the Philippines, and Cambodia, have sustained a growth rate of around and over 7 percent (Jennings, 2017) without experiencing significant slowdown in the past years.

However, the IMF's Regional Economic Outlook highlights future risks related to building resilience, ensuring sustainability, and guaranteeing fiscal support for the years to come; climate change, increasing prevalence of disasters, and demographic challenges all pose risks to sustained and equitable economic development (IMF, 2018). For example, in Nepal, future projections for growth rely heavily on the development of hydropower facilities across the country to support the increasing need for electricity output. However, much of the development is taking place in areas susceptible to the impacts of glacial lake outburst floods (GLOFs), which are projected to increase in frequency and severity because of glacier melt following climate change (Carrivick et al., 2016),

Disasters and climate-related hazards affect not only the economies in the region but the people as well. Socio-economic impacts of disasters can range from major losses of livelihoods (such as income-generating equipment) to the loss of homes, education, healthcare, and social welfare. In most of the countries,[3] disasters and hazards have usually affected healthcare facilities. This is a significant concern, especially in areas where local capacity is low and where remoteness does not allow rapid support to the communities. In countries such as Fiji, Papua New Guinea, and the Maldives, disasters affecting local healthcare units may contribute to increased mortality rates in the aftermath of events, especially for the elderly and children. Lack of local capacity also contributes to heightened vulnerabilities preceding disasters. For example, in PNG, the PNG, majority of the population residing in rural

DOI: 10.4324/9781003586869-2

regions have limited access to care facilities, meaning that ill or injured patients may have to travel for over a day to see their community's respective health services.

Frequent disasters causing loss of livelihoods, income generation equipment, homes, and natural habitats can lead to displacement and forced migration, which contributes to the growing urbanisation trend in Asia and the Pacific. For example, the loss of livestock because of a dzud in Mongolia has contributed to the domestic migration processes that are increasing the population of the Ger settlements outside of Ulaanbaatar. However, it has also been recognised that the availability of diverse opportunities and better standards for improved healthcare and education contribute to the attraction of domestic migration (International Organisation for Migration, 2018). Nevertheless, due to lack of space, people are inhabiting areas with a high risk of landslides and flash flooding, and the urban slums are often lacking basic infrastructure.

It should be noted that these impacts (and poverty) are often gendered; women and girls tend to be disadvantaged compared to men due to social normative practices and traditions which obstruct their access to education, land ownership, income generation equipment, employment opportunities, or social standing comparable to that of men. Furthermore, disasters tend to aggravate gender-based violence, which contributes to the stressors women and girls face in the aftermath of disasters. Elements of gendered vulnerabilities are present in all the assessed countries.

Climate Change and Its Impacts in Asia and the Pacific

Besides disaster risks, the countries in Asia and the Pacific are generally considered a high-risk area for climate change impacts, and the region will carry the most severe consequences of adverse weather, changing temperatures, ocean acidification, and environmental degradation. Risks are further exacerbated by the rapid pace of population growth, sometimes unplanned development of urban areas, and the countries' high reliance on the environment and ecological systems. The Global Climate Risk Index of 2019 explicitly highlights that the countries of South Asia are among the most vulnerable globally to the impacts of climate change, and five of the countries most affected by climate change between 1998 and 2017 are located there (GermanWatch, 2019).

Much of the climate change vulnerability of Asia and the Pacific stems from geography. Many of Asia's expanding cities and ports are located in the vicinity of the ocean and in a region exposed to a plethora of geophysical hazards, from typhoons to flooding, earthquakes, and tsunamis. Furthermore, the Pacific Islands are increasingly vulnerable given the low-lying topography, which leaves little room to escape the rising sea levels, and their freshwater aquifers – often resting as shallow lenses on the seawater – are extremely vulnerable to saltwater intrusion and droughts.

Also, climate change has been predicted to exacerbate the impacts of hydro-meteorological hazards, as increasing precipitation is likely to enhance flooding and storms due to increased regional precipitation. Conversely, rising temperatures and prolonged periods of drought may contribute to reduced water availability, and the decreasing volume of glaciers in the Himalayan region is likely to contribute further to the problem further. In Mongolia, the loss of permafrost has been linked to diminishing water resources, and the unusually high temperature rise has been found to increase the occurrence of steppe fires, as explored earlier.

Lack of safe water and food may also instigate increasing numbers of conflicts and contribute to existing ones. Also, loss of income may lead to forced migration, which is often concentrated in urban regions due to the availability of opportunities. However, this contributes to the problem of growing numbers of urban poor.

Disasters and Development at the Crossroad

The importance of harmonising the post-2015 development agenda into coherent plans and actions when facing the surmounting disaster and climate risks in Asia and the Pacific is an immense challenge. Given that the countries only have the upcoming decade to adequately prepare for the first catastrophic impacts of climate change (and less for unexpected disasters), streamlining efforts must take place now to guarantee that the understanding of risks is guided by both an understanding of disasters and climate and their interaction within complex systems characterised by diverse vulnerabilities of infrastructure and people.

The global community has adopted the Sustainable Development Goals (SDGs), the Sendai Framework for Disaster Risk Reduction (SFDRR), and the Paris Agreement to fight against climate change impacts. Optimising country actions towards achieving coherence of the three agendas has become another exclusive area to be pursued. The Intended Nationally Determined Contribution (INDC), submitted to UNFCCC, outlines climate actions by each country to reduce national emissions and adapt to the impacts of climate change (United Nations, 2019), and National Adaptation Plan is also a main instruments to identify linkages across policies and to create coherence across various climate and environment-related frameworks, with DRR synergies.

Given the high urgency of climate change and increased visible impacts of hydro-meteorological hazards, it is inevitable for risk-informed development to adopt more progressive climate-risk reduction and climate-sensitive DRR approaches. Some risk reduction measures simply transfer disaster risk in one area to another, while some might not be sufficient to tackle future risks of extreme climate events such as prolonged droughts and water shortages, intensified by the competing demand on water resources for development activities. Moreover, only investing in physical infrastructure, in the absence of environmental considerations, further undermines ecological services for risk reduction and climate change mitigation. Hence, integrated actions and futuristic solutions are crucial to address the underlying factors of people and sectoral vulnerability, combining environmental protection and minimizing the trade-offs between growth and sustainable development.

For example, in Nepal, the institutional foundation for climate change has traditionally been strong, and country has successfully piloted community adaptation programs through the Environment-Friendly Local Governance (EFLG) Framework and Local Adaptation Plan of Action (LAPA) in collaboration with the local government authorities (MoHA, 2017). Work in the area of climate change is reinvigorated with Nepal having ratified the Paris Agreement and now in the process of formulating National Adaptation Plans (NAP) to help address medium and long-term adaptation needs and reduce climate vulnerabilities through a sectoral approach (MoHA, 2017), focusing on agriculture, urban settlements and infrastructures, and climate-induced disasters (MoPE, 2017).

Despite the focus on economic-oriented development aspirations, disaster and climate risk are given due attention and are well-articulated in planning documents elsewhere as well. Pakistan 2025: One Nation – One Vision, fully aligned with the SDGs (Ministry of Planning Development and Reform, 2018) is taking into account climate change challenges as a priority for government action, and aspects with DRR-linkages are well noted as part of water security issue (ADB, 2017). Similarly, Mongolia aspires to be among the leading middle-income nations with a society consisting of middle and upper-middle-income classes by the year 2030, with a focus on preserving ecological balance (The Government of Mongolia, 2016). It is the issue of to what extent the planned actions can be realised at a practical level, and diversions from the plans are likely due to various development pressures, budget constraints, mounting risks, and the absence of a working modality for collaborative cooperation among different government functionaries.

Progress Made under Sendai Framework for Disaster Risk Reduction Priorities

Countries in Asia and the Pacific have demonstrated significant progress in Disaster Risk Management (DRM) following the conclusion of the Hyogo Framework for Action (HFA: 2005–2015), with continued efforts and upscaled actions to advance the implementation of the Sendai Framework for Disaster Risk Reduction (SFDRR 2015–2030). Major disaster events in the region – including Typhoon Yolanda in the Philippines in 2013, the Gorkha earthquake in Nepal in 2015, and the drought of 2016–2017 – 2017 – gave repeating wake up calls for the region to advance DRM beyond responsive action. The wealth of experience from recurrent disasters provides invaluable learnings and reaffirms the need to guarantee effective disaster risk management at all levels of government, in local communities, and in every sector of industry.

With the growing understanding of the development-disaster nexus, reducing vulnerabilities to disaster risks has gained greater recognition among government entities, CSOs, and development agencies. However, interventions are still limited even at the global level, and a baseline for good practices in terms of managing underlying vulnerabilities contributing to risks (such as poverty, environmental degradation, rapid urbanisation, and population growth) is still largely missing in hazard-prone areas (CRED, 2017). DRM work in various countries remains response-centric in practice. Despite such hindrances, policies focused on comprehensive Disaster Risk Reduction at all levels and across sectors have persistently gained momentum.

This section provides an overview of DRM regional progress and trends under the four SFDRR Priority Actions, based on the ten country reports. While the countries are immensely diverse in terms of hazards, environment, geophysical context, and vulnerabilities, as well as level of advancement in DRM, common outcomes being achieved are drawn and highlighted to identify regional challenges and priority areas for work in the future. This is done in recognition of the fact that combating the growing disaster and climate risks often requires joint efforts and streamlined transboundary initiatives.

Priority 1: Understanding Disaster Risk

To improve understanding of disaster risks, risk assessments have been carried out consistently in all countries during the past decades, mainly focussing on the physical aspects of hazards. Many are also outdated. Furthermore, they rely on project-based exercises for academic purposes or small-scale projects in particular locations, conducted by specialised technical teams employing various technical tools. Thus, a comprehensive, commonly referenced baseline and understanding for conducting risk assessments are missing. Factoring in the vulnerable, exposed attributes of the population and systems into risk computation methodologies requires more comprehensive baseline data. Discrepancies also remain between the scientific-based assessment results (employing probabilistic and modern complex risk modelling) and optimizing their use for end-users, development planners, and local authorities. Perhaps, to complement the nationwide risk assessments and those conducted by external stakeholders, a simple but technically sound methodology, suitable in terms of the technical competency required, should be developed to guarantee that risk assessment exercises are practical for local DM authorities and stakeholders.

At the country level, many nations have started developing systematic data platforms, many of which utilise advanced GIS technology. Furthermore, information hubs for hosting geospatial data covering hazards and risks, alongside applications for managing comprehensive disaster information, have been developed to a varying degree.

Priority 2: Strengthening Disaster Risk Governance to Manage Disaster Risk

One of the key achievements across the assessed countries is the establishment of robust risk governance systems, supported by solid legislative frameworks and regulatory provisions. Institutional arrangements for DRM have often been set up for each respective level of administration, with defined mandates and continued institutional capacity strengthening. However, when looking back, DRM has not been among the core mandates of government institutions, rather it has been auxiliary to either the army or internal affairs ministries. The performance of emerging DRM structures remains intermittent due to fluctuating frameworks, overlapping responsibilities, changes in leadership, transitions of staff, and/or shortages of resources and budget to operate. Greater efforts are required to increase the recognition and profile of disaster and climate risk management across institutional entities.

As of now, many countries revitalised their policy frameworks and plans, addressing the spectrum of DRM with defined country-specific goals, in line with the SFDRR progress monitoring indicators and reporting timeline. Some of the most up-to-date policies and plans are the National Disaster Risk Management Plan of Sri Lanka (2018–2030, draught), Fiji's National Disaster Risk Reduction Policy (2018), Pakistan National Disaster Management Plan and Implementation Road Map (2016–2030) and Nepal National Disaster Risk Reduction Policy (2018), which comprehensively and proactively seek to address disaster (and sometimes, climate) risks at all levels of government.

However, procedural frameworks for operationalising DRM across multi-tier risk governance structures and across multi-sectoral government functionaries and stakeholders are yet to be established in many of these countries. Coordination mechanisms, platforms for inter-agency collaboration, and Standard Operating Procedures (SoPs) are essential to move towards integrated, well-coordinated, and comprehensive DRM.

Recent progress in the region highlights country commitments to increasing progressive risk governance, focussing on capacity building at various levels, and creating robust mechanisms for guaranteeing effective action. Sri Lanka has started the process of improving the capacity of local-level operators by institutionalizing local-level support in the SLCDMP planning (Ministry of Disaster Management, 2014). It has been developed in accordance with the National Disaster Management Plan of 2013–2017 that further mandated a multi-sector approach under which all sub-national levels of administration, NGOs, and grass-root level organisations must have their operational plans, harmonised with the national DM plan to translate policies into tangible action (DMC, 2019). Similarly, in Pakistan, institutional capacity strengthening has taken place for the capacity building of disaster management bodies at the Federal, Provincial, and District levels, with support by the World Bank. Such progress is evident in many of the countries, but the localization agenda of disaster and climate-risk reduction efforts should be increasingly mainstreamed throughout Asia and the Pacific.

Priority 3: Investing in Disaster Risk Reduction for Resilience

Investing in disaster risk reduction is elemental for the countries in the region, alongside investments in climate resilience and adaptation. This section highlights the current progress and challenges related to managing disaster finances in the assessed countries.

Under this priority, large-scale investment for flood risk reduction is ongoing in various countries, including structural flood mitigation projects for flood protection, irrigation channels, small dams, and protective works in Pakistan, implemented annually under the auspices of the FFC and the provincial irrigation departments (Sayed & González, 2014), and Flood risk mitigation structures for the urban development and physical regeneration of Metro Colombo City in Sri Lanka. Besides such large investments, initiatives to integrate DRR for safer and more resilient sector development have emerged in recent years. The Ministry of Planning and Investment (MPI), Government of Lao PDR, with support by the World Bank, has made climate change and disaster risk considerations mandatory in the public investment review process since 2017 (GFDRR, 2019), which has evolved from various attempts to integrate disaster and climate risk considerations into development decision-making and sectoral planning. Demonstrative projects were carried out in 2012–2016 in irrigation schemes, landslide mitigation in road infrastructures, piloting the use of DRR-inclusive appraisal criteria for development projects, and policy guidelines for risk-inclusive Lao PDR's 8th National Socio-Economic Development Plan (NSEDP) 2016–2020 (ADPC, Resilient investment decisions in Lao PDR: Impact Story, 2016).

Countries in the region have also put mechanisms in place to enhance social welfare and social protection for different vulnerable groups, aiming to achieve vulnerability reduction and poverty alleviation. For example, Mongolia has established a social welfare system, which consists of 71 different programmes managed by the Ministry of Population Development and Social Protection, covering social pensions and cash transfers to citizens with special needs, such as people with disabilities or those who lack family connections, and provides child support to poorer households. Similarly, Pakistan has initiated social safety net schemes for labour force and non-labour forces, including Workers Welfare Fund, the Employees Old-Age Benefits Institution, the Zakat Fund, and the Benazir Income Support Programme (BISP). Established in 2008, BISP helps reduce social and economic vulnerability, covering more than 7.7 million households (ADB, 2019).

Priority 4: Enhancing Disaster Preparedness for Effective Response to 'Build Back Better' in Recovery, Rehabilitation, and Reconstruction

To begin with early warnings, tangible progress has been achieved in terms of developing weather monitoring services at the country level. Data interfaces and communication networks are increasingly linked with regional and worldwide meteorological monitoring systems, owing to extensive technical support and institutional capacity building from various development partners, in particular the World Meteorological Organisation (WMO), UNDP, and the World Bank. To a certain extent, this has helped in establishing standardised meteorological monitoring systems and services across the region, which have significantly improved short, medium, and long-range climate forecasting and supported the development of robust early warning systems, especially for flooding and storms. As a result, public access to real-time hazard monitoring, situation reports, and early warnings has significantly improved as well.

In Fiji, the Fiji Meteorological Service (FMS), under the Ministry of Disaster Management and Meteorological Services, provides early warnings for cyclones, rain and droughts, and disseminates the information through SMS, social media, as well as radio and TV networks. In the Philippines, early warning systems are further enhanced by the Free Mobile Disaster Alert Act of 2014, which mandates telecommunication operators to issue free public warnings via mobile phones, as required by the state, PHIVOLC, or PAGASA (GoP, 2014). In Mongolia, National Emergency Management Agency (NEMA) is able to disseminate warning messages easily to provinces and soum (district) levels; however, it is difficult to reach remote herder communities despite the far-reaching coverage of the current initiatives (CFE-DM, 2018). The latterly mentioned challenge is shared by many countries characterised by remoteness and geographical isolation; Fiji, Maldives, and Papua New Guinea have been struggling to disseminate early warnings to the lowest levels (communities and households without access to usual EWS channels or services) in the past.

On 'building back better' in recovery and reconstruction, post-disaster needs assessments (PDNAs) have become compulsory exercises in the aftermath of major-scale disasters, guided by technical teams, and mobilised with the support of development partners, including UNDP, the World Bank, and GFDRR. On the ground, work is usually conducted by field assessment teams comprising officials of line sectoral agencies, local governments, and volunteers, with National Disaster Management Offices (NDMOs) in charge of the coordination. However, the extent and quality of the PDNAs often vary; it depends on the local capacities, complexity of disasters, scale of impacts and technical know-how, among other factors. Also, the methodology developed for PDNAs remains to be contextualised according to key sectors and cross-cutting themes in many of the countries due to certain limitations such as lack of baseline data, sector-specific data, and local technical viability. Even if the understanding of disasters has been attained, transforming the lessons learnt to practical implementation is difficult again due to lack of capacities and budget. In the years following Cyclone Winston affecting Fiji in 2016, recovery and reconstruction reflect that risk integration and 'building back better' approaches and understanding of how shared experience from humanitarian response could be used to inform development are difficult to mainstream at the sub-national levels (Winterford & Gero, 2018).

Role of Stakeholders in Asia and the Pacific

Given the challenges related to a lack of local capacities, limited budgets, resources, and otherwise complex, diverse settings, countries may sometimes struggle to cope with the demands and requirements of managing disaster and climate risks internally. Thus, the role of stakeholders is, and has been, important role since technical capacity and donor-driven activities can create a significant difference, especially for local communities and sub-national levels.

In recent years, roles of a wider range of stakeholders – including community self-help groups, women-led groups, the private sector, as well as media – have emerged and become important actors in the disaster risk management domain. Moreover, conventional stakeholders such as development partners, INGOs, NGOs, and Community-Based Organisations (CBOs) have further enhanced their engagement and contributions. UNDP, WMO, and the World Bank remain key partners working intensively with national governments on DRM technical and institutional strengthening. Financial support from donors, including USAID, DFAT, the European Union (EU), GIZ, UNDP, ADB, the World Bank, and JICA, has been instrumental for country implementation of DRM interventions in Asia and the Pacific.

For humanitarian assistance, the Red Cross Society and Humanitarian Country Team, led by UN agencies and the inter-agency cluster-based response system, serve as the main mechanisms that coordinates and complement the emergency response system of the government. In PNG, emerging challenges in coordination and communication during the El Niño drought response in 2016 resulted in the formation of the Highlands Humanitarian Hub (HHH), a multi-stakeholder body comprising Provincial Disaster Coordination Committees, NGOs, churches, the Red Cross, and multilateral organisations. The HHH was established exclusively to facilitate coordination mechanisms in the Highlands region, with a particular focus on food security issues. Its scope of operation later expanded to all key emergency clusters (HAG, 2018).

Creating platforms to organise non-state DRM actors into unified group has evolved in different countries. Among these are the Cambodian Humanitarian Forum, set up in 2012 and comprising more than 120 NGOs to strengthen the leadership capacity of NGOs (CHF Cambodia, 2019) and the Disaster Risk Reduction Network Philippines, composed of over fifty organisations with expertise is focused on different pillars of disaster risk reduction. The localisation of emergency response and preparedness has called for intensified intervention to synchronise capacity building of local NGOs, enhance local coordination mechanisms, and synergies between INGOs and local NGO networks, thereby ensuring self-sustaining capacity (ADPC, Uniting local NGOs against disasters in Cambodia, 2016). The private sector has also gained recognition and a high profile in DRR with the Philippine Disaster Resilience Foundation (PDRF) established to envision a strong, resilient, and disaster-ready Philippines, and PDRF's Emergency Operations Centre (EOC), the first national private sector EOC, operating since April 2018 (PDRF, 2019).

Besides assuming roles in supporting the government in disaster relief and response, the private sector's role in making sectors resilient to disaster and climate risk has become increasingly crucial. In the Philippines, private investment driving infrastructure, public utilities, and industrial development in the country could significantly reconfigure disaster risk. Disaster risk-informed investment under the Private Sector Alliance for Disaster Resilient Societies (ARISE), launched in 2015 and led by the United Nations Office for Disaster Risk Reduction (UNISDR), has urged more members of the private sector to collaborate with local government units to raise awareness and deliver impacts on DRR management and resilience.

Regional platforms play a unique role in creating and sustaining DRM momentum and partnerships, providing technical support to advance country DRM, and promoting regional synergies. Regional entities, mechanisms, and agreements in which countries actively take part include Mekong River Commission (MRC), the ASEAN Agreement on Disaster Management and Emergency Response (AADMER) – a legally binding agreement for the 10 member states of ASEAN, the ESCAP/WMO Typhoon Committee, to name a few. In the Pacific region, the Secretariat of the Pacific Community Applied Geoscience and Technology Division (SPC/SOPAC) serves as a Pacific regional centre that supports region-wide DRM initiatives and provides long-term contributions to advance DRM in individual countries in the Pacific, including Papua New Guinea. The Secretariat also supported Pacific Island countries in global negotiations including finance issues at the UNFCCC COP.

Challenges and Future Priorities for Action

Most of the countries in Asia and the Pacific are still in the process of rapid growth and development in terms of managing their internal affairs; challenges persist in attaining an equilibrium between competing development demands such as the needs for healthcare, education, rural livelihoods, employment opportunities, and pursuit of economic development, all the while facing increasing risks in the region. Striving to achieve this balance among different sectoral endeavours to ensure long-term sustainability in economic growth, social development, and environmental well-being is admittedly difficult.

To add to the existing stressors, the multi-faceted relationship between disaster and climate risks either in urbanised setting, remote rural contexts, or growing secondary cities, has manifested into chronic problems desperately awaiting desperately for long-term solutions. These include worsening water shortages due to droughts or groundwater salinization, coupled with increasing demand for industrialisation and agriculture, and limited access to once shared resources due to privatisation, which limits livelihood opportunities. Furthermore, the loss of livelihoods, mainly in subsistence agriculture, to disasters and adverse weather has driven people to expand settlements to unsafe land or caused them to resort to negative coping mechanisms, further contributing to environmental degradation. This may also lead to difficult-to-manage expansion of urban slums unless alternative livelihood options are provided for rural populations. In certain circumstances, these processes may create interlinked acute shocks triggered by hazards, causing severe damage to existing systems. For example, new cities in the region tend to suffer from worsened urban flooding, which further damages city lifelines and/or core functions that provide support to businesses and economic activities. Thus, seemingly simple hazards can rapidly turn into cascading events.

Countries also struggle with the paradox between sustainable development and rapid economic growth, often built on extractive industries. In Mongolia, the mining industry and investment in infrastructure have spurred economic performance, while contributing to increasing pressures of climate change, poverty, and disaster risks. Soil erosion and dust storms have been established to be more severe at mining sites, and the privatisation of land has started to landlock herders from accessing previously commonly shared areas.

To centralise limited resources, the government of Maldives is implementing an ongoing plan to consolidate the population from outer islands to the Greater Male area and nearby islands in hopes that it could reconfigure disasters and climate risks through increased efficiency and planning. However, infrastructure and increased density of the built environment could result in increased vulnerability to floods, while waste and its disposal are likely to exacerbate water pollution and health problems during flood-affected periods without effective waste management. In Lao PDR, land concession policy to attract foreign investment might unintentionally increase the stress on land, water, the environment, and ecological systems, which offer unique ecosystem-based services for DRR and act as an important carbon sink. The interface of these socio-economic, environmental, and human induced factors must be addressed in a more holistic manner, with durable solutions that offer new opportunities for community development, empowerment of the people, and social transformation.

Future Priorities

When disaster and climate risk management are taking place in a setting where risks are exacerbated and interlinked with multiple, complex functions – such as the severity and frequency of hazards and climate uncertainty – investing to strengthen risk governance itself is crucial. Investments and resources to capacitate action and technical know-how of DRM officials at national and sub-national levels should be among the highest priorities for countries in Asia and the Pacific. An extensive review of existing capacities within NDMOs vis-à-vis the level of technical expertise required for their work is needed to identify needs and gaps for institutional capacity improvement.

Furthermore, budget deficits often adding to the challenges obstructing sub-national level development and implementation of risk reduction activities. In many countries, DRR and CCA initiatives rely on financial support from international donors and development partners to meet financial gaps. Addressing the lack of DRM resources must be done in a more systematic manner to move away from donor dependency to secured localised funding in terms of risk financing. Good examples of national efforts to improve investments in disaster risk management include Pakistan, which received US$ 200 million loan approved by the ADB in 2016 to support the establishment of a National Disaster Risk Management Fund. The World Bank has supported the establishment of Disaster Risk Financing strategies and DRM funds at the provincial level (ADB, 2019). Similarly, strengthening financial resilience has become a priority area for Sri Lanka. The establishment of a Development Policy Loan (DPL) with a Catastrophe Deferred Draw-Down Option (or CAT DDO), with support by the World Bank following the 2016 floods and landslides, has led to the exploration of different disaster risk financing options to better cope with the fiscal burden on the government for contingency spending, compensation to affected populations, and advancing mainstreaming DRR into various sectors (World Bank, 2014).

Related to finance and prioritization, the availability of disaster and climate-related data, and its management (especially in terms of SADD data availability) must be met with systematic monitoring and evaluation of country's operationalization of the SFDRR. However, technical challenges are limiting the usage, and the lack of local capacity, technical skills, and resources limits the objective to increase understanding of disaster and climate impacts. Thus, building local capacity, and harmonising sub-national level governments' approaches to data management should be among the first priorities to further safeguard countries from the impacts of climate and disaster by thoroughly understanding risk. It is also required for the prioritization of initiatives, especially in a context where resources are otherwise low.

Finally, focus should also be directed towards managing vulnerable systems as interlinked, diverse arrangements of moving pieces (environment, population, hazards, climate, and physical infrastructure) which are interlinked in a way where hazards may trigger complex disasters. Understanding the needs of different sectors, and the effects of their development – such as environmental degradation, increasing urban risks, and cascading events – must be done from a holistic perspective to further the understanding of vulnerabilities. Poverty reduction is an essential element of this discourse – often, the most severe impacts of disasters are borne by those with no access to risk transfers, and when their reliance on vulnerable sectors is high. Efforts to identify how to better manage the future displacement and forced migration caused by the loss of agricultural output to low-income farmers, for example, must be made well before the actual risks are materialising on the ground. Similarly, environmental impact assessments and identifying high-risk areas for environmental degradation should be led with systematic approaches to guarantee that the trade-offs between sustainable development and growth are minimised and balanced to prevent any further degradation of natural systems.

Notes

1 Asian Disaster Preparedness Centre (ADPC), Thailand aslam@adpc.net
2 Pridi Banomyong International College, Thammasat University, Thailand Shweta .s@pbic.tu.ac.th
3 We have to mention the names of the countries assessed.

References

ADB. (2017). *Climate change profile of Pakistan.* https://www.adb.org/sites/default/files/publication/357876/climate-change-profile-pakistan.pdf

ADB. (2019). *Diagnostic on the current availability and usage of insurance, reinsurance, and capital markets for disaster risk financing.* https://www.adb.org/sites/default/files/publication/483686/pakistan-disaster-risk-financing-assessment.pdf

ADPC. (2016a). *Resilient investment decisions in Lao PDR: Impact story.* http://www.adpc.net/igo/category/ID1073/doc/2016-swn0Re-ADPC-Impact_Story_-_Lao_PDR_Final_-_Web.pdf

ADPC. (2016b). *Uniting local NGOs against disasters in Cambodia.* https://www.adpc.net/igo/category/ID1108/doc/2016-lpg6Jw-ADPC-Impact_Story_-_Cambodia_Final.pdf

Carrivick, Jonathan L. and TWEED, Fiona (2016) A global assessment of the societal impacts of glacier outburst floods. Global and Planetary Change, 144, 1–16. ISSN 09218181

CFE-DM. (2018). *Mongolia: Disaster management reference handbook* (1st ed.). Center for Excellence in Disaster Management and Humanitarian Assistance.

CHF Cambodia. (2019). *Cambodian Humanitarian Forum (CHF).* http://chfcambodia.net/what-we-do/

CRED, E.-D. a. (2017). *Economic losses, poverty & disaster 1998–2017.* https://www.preventionweb.net/files/61119_credeconomiclosses.pdf

DMC. (2019). *Disaster management centre: About us – National disaster management plan.* http://www.dmc.gov.lk/index.php?option=com_content&view=article&id=74&Itemid=236&lang=en#

GermanWatch. (2019). *Global climate risk index 2019 who suffers most from extreme weather events? Weather-related loss events in 2017 and 1998 to 2017.*

GFDRR. (2019). *Lao people's democratic republic.* https://www.gfdrr.org/en/lao-peoples-democratic-republic

GoP. (2014). *Republic Act No. 10639.*

Government of the Republic of Fiji. (2018). *Republic of Fiji national adaptation plan: A pathway to climate resilience.*

HAG. (2018). *Extractives and emergencies: The Papua new guinea earthquake response.*

IMF. (2018). *World economic and financial surveys regional economic outlook.*

International Organization for Migration. (2018). *Mongolia: Urban migrant vulnerability assessment.*
Jennings, R. (2017, March 23). These are the fastest growing economies in East Asia in 2017. *Forbes.*
Ministry of Disaster Management. (2014). *Sri Lanka comprehensive disaster management programme 2014–2018.*
Ministry of Planning Development and Reform. (2018). *Overview of SDGs.* https://www.pc.gov.pk/web/sdg/sdgpak
MoHA. (2017). *Nepal disaster report 2017: Road to Sendai.* http://drrportal.gov.np/uploads/document/1321.pdf
MoPE. (2017). *Vulnerability and risk assessment framework and indicators for National Adaptation Plan (NAP) formulation process in Nepal.*
PDRF. (2019). *About PDRF.* https://www.pdrf.org/about/letter-from-the-president/
Romel, V., & Reed, J. (2019, March 26). The Asian century is set to begin. *Financial Times.*
Sayed, A. S., & González, P. A. (2014). Flood disaster profile of Pakistan: A review. *Science Journal of Public Health,* 2(3), 144–149. https://doi.org/10.11648/j.sjph.20140203.11
The Government of Mongolia. (2016). *Mongolia sustainable development vision 2030.*
United Nations. (2019). *United Nations framework convention on climate change.* https://unfccc.int/process-and-meetings/the-paris-agreement/nationally-determined-contributions-ndcs#eq-2
Winterford, K. and Gero, A. (2018). Humanitarian response for development in Fiji: lessons from Tropical Cyclone Winston. IIED, London.
World Bank. (2014). *Advancing climate and disaster resilience in Sri Lanka.* https://blogs.worldbank.org/endpovertyinsouthasia/advancing-climate-and-disaster-resilience-sri-lanka

2 Geopolitics of Climate Change

Alexandre Lambert

Introduction: From 'Conserving the Nature' Towards 'Securing the Climate'

For decades, environmental protection was focused on *natural ecosystems* and *biodiversity*, while the 'climate' was largely a matter addressed through the lens of meteorology. Thus, environmental activities were about the protection and preservation of healthy living conditions for all humans and non-human species, including plants, animals, and microorganisms. The focus was on protecting the *biosphere*, given also that our planet effectively loses an ever-increasing number of living species every year forever.[1] In this context, the anthropogenic causality for the loss of our planet's biodiversity has arguably been less controversial than humankind's contribution to global warming. Anyway, proactive environmental protection and advocacy were to provide for *public goods*, including in terms of public and global health, and with human health gradually extended to include non-human species. For instance, the concept of '*One-Health*'[2] correlates the natural environment with both human and animal health – a concept that becomes relevant also to the sustainable management of agricultural and food systems. Finally, the idea of One Health has been further elaborated to include all living species and both human and non-human ecosystems, with global health being ideally transferred ideally into an ecological all-in-one concept now referred to as *Planetary Health*, and for which *The Lancet* even issued an open-source series in 2016.[3]

At the same time, an assertive and increasingly powerful public discourse on 'climate change' has emerged, with a tendency to quasi-hijack the global ecological agenda, reducing the biosphere to the *atmosphere* with its typical focus on gas molecules, especially CO_2, framed as if it is an 'existential threat' to humankind, although it is a vital gas for most plants. This trend of '*environmental reductionism*' or '*environmental diversion*' can also be observed in the context of some international environmental NGOs, who are increasingly lobbied by the for-profit private sector that tends to use 'climate environmentalism' for corporate branding and to eventually circumvent environmental accountability. Consequently, international ecological discourses experience a 'cognitive compression', whereby *environmental*

DOI: 10.4324/9781003586869-3

protection is being recycled into *'climate* protection', and in the context of which *natural conservation* tends to be framed as *'climate conservation'* – with the variation of average annual atmospheric degrees becoming a priority item on many government and non-governmental agendas around the world.

When looking closer at this puzzle, this phenomenon almost looks like a 'corporate trick' orchestrated from within the financial industry. With the predominant environmental focus now being placed on carbon-based / fossil fuel emissions and the so-called Greenhouse Gases (GHG) perceived to cause 'global warming', instead of being held accountable for their overall ecological footprint and related negative environmental and social externalities, strategic industries and financial markets now seem to even 'benefit' from climate change, as CO_2 is effectively *commodified*, offering a new multi-billion-dollar business opportunity. For instance, *carbon-trading*, with new business ventures spreading, is becoming a highly profitable endeavour in financial markets and stock exchanges. And since strategic industries are never too far away from national security communities, an adjacent *securitisation of the climate* gradually sidelines genuine environmental protection. Even pharmaceutical industries and psychiatrists encounter climate change-related public health concerns and therewith new medical industry and business opportunities in view of an emerging mental health disorder reportedly affecting an increasing number of youth populations around the world: *'climate* anxiety'.

Climate Change Becoming a Major National Security Concern According to Mainly Western Narratives

Geostrategic communities increasingly frame climate change as if it had become a major threat to national and international *security* and political stability. Former director of the CIA, General David Petraeus, considers climate change a 'threat multiplier' that causes new 'geopolitical stressors', and with reference to the Biden Administration's national security advisership, he adds that there is now some agreement within the US government to address climate change as an 'urgent *national* security priority'.[4] Adding, however, that the costs and challenges of climate change and global warming are likely to hurt *developing countries* disproportionately, e.g., by increasing already existing pressures on resource scarcity, including access to fresh water, as well as poverty, corruption, and 'authoritarianism', the resulting implication and main policy response for the United States, according to Petraeus, is 'to reinforce our (US) defences and batten down the hatches'.[5] Meanwhile, the prestigious British consulting firm PriceWaterHouseCoopers (PWC) identifies climate change among the *Five Megatrends*, with its corresponding report's subtitle making no mistake about the fact that this has got implications for '*global defense and security*'.[6] Herewith, the lead Trans-Atlantic powers, both from a geopolitical and geoeconomic perspective, consider climate change not only a major security issue, but also a new priority for *national defence* on a planetary scale. In other terms, as Shane Tomlinson conjectures, climate change has not only become a *'major shaper of geopolitics'*, but also a *'major driver shaping the financial system'* and thereby becoming a *'major fixture'* of the G7, G20, UN agencies, and the International Financial Institutions (IFIs).[7]

The increasingly visible nexus between climate change and national security can also be highlighted in view of the following socio-political assertion: *'The climate is not a conflict protagonist, but protagonists may exploit climate'*.[8] With the corresponding book chapter title (no. 2.5.2) in his study on *The Geopolitics of Climate Change*, Peter Halden, already back in 2007, illustrated how Western strategic communities have tended to *securitise the climate itself*. Issued in December 2007, the Swedish study was issued in the heat-up to the *Great Financial Crises (GFC)* and corresponding meltdown at Wall Street, with Lehman Brothers collapsing less than a year later in October 2008. Halden's prognosis was that due to climate change, a *quasi*-generic syndrome would spread around the world that his report calls the *'loss of state-interactive capacity'*.[9] Accordingly, and 'judging by the purpose of the IPCC',[10] as Halden formulated, 'climate change would negatively impact *entire regions* more than individual countries', adding, however, that (regional) *states' individual preparedness* would vary.[11] And in the case of comparatively vulnerable regions such as for instance, *South Asia*, Halden's assessment was that their 'security dynamics appear to be endogenous to a high extent'.[12]

Interestingly, Halden would consider the *'North-Atlantic security community'* (thus: NATO) as the *most important 'group' of countries*[13] in response to climate change, not only, according to Halden, given this group's major role in foreign aid and Foreign Direct Investment (FDI), but also given its advanced (and implicitly military) capacity for *'international humanitarian intervention'*, which in turn corresponds with his above-stated assertion that losses in state-interactive capacity could well lead to *'regional conflicts'*. Coincidence or not, Halden's report in Sweden timely came with NATO's increasingly assertive outreach to the Arctic region and corresponding interest in strategically associating Scandinavian countries – perhaps anticipating what now suddenly becomes a concrete scenario in the wake of the war in Ukraine. It also coincides with both the start of the GFC (2007–2009) and the International Energy Agency's (IEA) *World Energy Outlook 2007* underlining *i.a.* the increasing fossil fuel use, especially of *China and India*, and arguing that if no 'urgent action' is taken to check global warming accordingly, global temperatures would reach a *quasi*-irreversible level of 3% above pre-industrial levels.[14]

What is, however, symptomatic about Western strategic reports on climate change is that they hardly provide concrete, namely comparative, international data on *both absolute and per capita* GHG emissions. In fact, the group of G7, as well as most NATO countries along with some of the West's global strategic partners, remain clearly among the world's top GHG emitters. While the five biggest *absolute* emitters do indeed include China and India, next to the USA, the EU, and Japan,[15] in terms of *per capita* emissions, Canada, Australia, and the USA figure among the top-10, with Estonia, the Czech Republic, the Netherlands, Japan, and Germany coming under the world's top-20. Moreover, if certain Western allies, e.g., among the group of the Gulf Cooperation Council (GCC),[16] would be added to this picture, this combined strategic cluster of countries would represent a significant and upper share of the world's top-20 *per capita* emitters. In contrast, the Russian Federation (just world rank no. 16) is the only BRICS country figuring among the (*per capita*) top-20.[17] Accordingly, the *per capita* GHG emissions of both China and India remain significantly lower than the average of both OECD- and former Soviet Union countries, in other terms: the *industrialised Global North*.

Thus, when looking at the global spectrum of distributive GHG emissions, it becomes clear why most countries in the Global South continue claiming their *legitimate right to economic development, including industrialisation*, and why they also argue that if anthropogenic climate change and global warming exist, it has largely been caused by the GHG emissions of established industrialised nations for the past two centuries.

Scientific Climate Facts and Figures vs. Neo-Religious Belief Cultures

Julius Caesar once proclaimed that 'Men readily believe what they *want* to believe' (In Latin: *Fere libenter homines id quod volunt credunt*). With Bill Bonner, this lesson from the legendary Roman Emperor also suggests that 'investors will believe what they *want*'.[18] In our context, this means that financial capitalism is unlikely to protect the natural environment.

Anyhow, for lead protagonists of climate change, global warming is represented as a *quasi*-absolute truth, with their global campaign resembling ontologically what US President Bush Jr. reportedly proclaimed (including to US allies) in the wake of 9/11 in view of the unlashed 'Global War on Terror' (GWOT): 'you are either with us (US government), or you are *against* us!' Coming as an implicit geopolitical warning, the pledge would also be addressed to US citizens themselves: if one did not endorse the 'official version' according to the US government, e.g., about what happened on 9/11, one wasn't considered a 'patriot', coinciding with the *Patriot Act*, which curtailed civil liberties in the USA.[19]

In our context, within and beyond (climate) science, to claim that something is 'absolutely true' will rarely convince all those who prefer to think that the truth is hardly ever about 'black or white'. For instance, some of those thinking in more differentiated terms claim that there might be a *third scenario*, namely that climate change and global warming would occur indeed (human-made or not), but that it wasn't dangerous for the Planet.[20] In contrast, hard advocates of climate change and global warming do not only proclaim apocalyptic scenarios for humankind – unless, as they warn, 'urgent measures' are taken now! They also pretend, as indicated, that respective scientific models remain on quasi 100% evidence, something held to be positively verifiable by worldwide peer-review, and that all antitheses remain not only untrue but also 'scientifically misconceived'.[21] Herewith, climate change along with the anthropogenic causality of global warming is being portrayed as if it was a 'given fact', almost like a *(para-)religious* belief or doctrine. In geopolitical terms, any *doctrine* will most likely reflect *societal power dynamics*. In our context, the increasing power of contemporary *'climate scientists'* arguably compares to the scientific leadership that 'Jesuits' once exerted during medieval and early modern Christendom.

Notably, during the late Middle-Ages in Europe, there was an impressive and heated *academic* debate with the aim of eventually 'proving' the existence of God. However, anything that resulted from this debate, let alone in terms of scientific evidence and coherence, was that if God existed, it was merely in the imagination of people. So, God's existence kept being scientifically unprovable, essentially since religious faith in God isn't about 'facts', but rather a matter of 'belief'. Nonetheless, some critical voices of climate science refer to certain flaws in *research methodology*, including the way key documents are *cited* in the scientific literature. For instance, Gavrilov et al. point out with reference to the *IPCC Assessment Reports* that those were not even themselves having the status of scientific papers, which in turn contrasts with their almost exclusive status as the most cited reference documents on climate science, to the extent that today, 'a scientific paper on this topic cannot be published without citing some of these documents' – which in turn entails that their status can indeed be compared to (faith-based) 'holy books'.[22]

Comments like these entail that there might be a 'political', if not 'geopolitical' agenda behind a (perceived or tacit) *doctrinal truth* about climate change. According to Gavrilov et al., not only could 'an "uncritical acceptance" of the doctrine of global warming ... have potentially detrimental implications for the *security* of countries and people'; it is even asserted that the climate change movement may actually be 'instigated by the most influential states ... (not only for) the "quest for profits' (*quasi* an "agenda behind") ... (but also for the sake of) ... *keeping the current relations of geopolitical forces in the world intact*'.[23] Perhaps, along with the authors' corresponding view that climate-related research should be brought back to an *"actual scientific* framework," this might in turn, be blamed and accused as a 'conspiracy theory' – just like 'non-believers (in God)' would once be persecuted. Anyway, certain parallels indeed appear between *monotheist* religious faith and *monolithic* scientific doctrine.

A deeper look into the *de facto politico-scientific debate* on climate change shows that the spectrum of *sceptical* or more *differentiated* perspectives on climate change remains remarkably broad and diverse. For instance, this is illustrated in an assessment of climate-related arguments brought up by conservative think tanks in the USA and Germany. The list below is but a few examples related to corresponding politico-ideological deliberations on climate change with potentially geopolitical implications:

- hydraulic fracturing (to extract natural gas) being 'better for the climate'[24]
- the United States not being themselves affected by environmental refugees[25]
- climate advocates and (so-called) 'environmentalists' representing unelected bureaucrats attempting to grab power and micromanage the deindustrialisation of the economy and the shutting down of market capitalism
- Paris Agreement being a blank check for China and India to increase their emissions
- international climate mitigation response being incoherent, with industrialised nations such as for instance, Germany relocating GHG emissions abroad.[26]

Anthropocentric Solutions Won't Compensate for the Preferred Focus on Anthropogenic Causality

One of the main quarrel apples of the politico-scientific debate on climate change and global warming is whether it is fully or just partially human-made. However, there is also another branch of the controversy pointing out that the problem is not only – or not even primarily – about the perceived *anthropogenic* origin of the phenomenon but also about the predominantly *anthropocentric* response to it. Ericka Centeno underlines how culturally and socio-politically embedded mechanisms have been shaping the climate change *agenda*, and how *cognitively*, this policy agenda suffers from '*physical reductionism*'[27] in which other living species other than humans play almost no role, and where solutions and responses are framed by mainly mercantilist, utilitarian, and technocratic approaches. And while *systemically*, it effectively *reproduces* 'an agenda aimed at protecting the current socio-economic system', including gender-stereotyping,[28] *normatively* (e.g., via climate-related negotiations), it is guided by '*geopolitical tensions* facilitating the control of rich countries over less developed countries'.[29] Arguing from the perspective of *political ecology*, Centeno further emphasises that climate change discourses tend to *blur social aspects*, and that proposals for the climate agenda referring to *social justice* are frequently overlooked. In fact, such proposals include highly strategic aspects such as the:

- Identification of sectors and agents who are responsible for the highest emissions;
- Exclusion from the carbon market of economic activities, including extractive practices, that externalise social and ecological costs.[30]

Physical reductionism and anthropocentric foci that tend to '*instrumentalize nature*' are also highlighted in the tendency within climate change discourses to reduce the *biosphere* to the *atmosphere*. For instance, in the context of discussing industry's *'negative externalities'* on the environment and human-induced *'tragedy of the commons'*, narratives go as far as to portray the atmosphere as 'being exploited and depleted'[31] – as if gas molecules were living species or precious stones! In contrast, most studies fail to highlight the effects of GHG emissions *beyond CO2*, let alone specific gases and other particles emitted by industries throughout the world that are rather toxic for both human and non-human health. This reductionist lens on *environmental pollution* potentially hides *geopolitically motivated agendas*, suggesting in turn, as indicated above, to 'secure' if not 'defend' the climate – short of actual environmental (*vs.* atmospheric) protective measures. Herewith, the new focus on 'climate talks' deviates the attention from 'talking about the environment', as the 'climate' is *de facto* 'weaponized', e.g., against the background of new great power rivalries. Furthermore, the *geopolitics of climate change* also reflects a *Neo-Cold War* game, in which two opposing camps confront each other economically, politically, and ideologically, both in the international and domestic realms.

A Neo-Cold War Syndrome with Inverse Politico-Ideological Patterns

Thus, well beyond the scientific framework, political deliberations and public debates on climate change increasingly resemble a *'Cold War 2.0'* dichotomy between roughly a new political 'left' versus a new political 'right', in the context of which the 'new left' can be associated with *climate change 'believers'*, while *climate change 'deniers'* tend to be associated with the 'new right'. What is interesting about this *new ideological polarisation, especially* as manifest especially in Western societies, is that in contrast to the old Cold War that ended in 1991, contemporary ideological patterns are structured almost inversely. During the old ideological antagonism, in the West, it was the leadership of the political right claiming a legitimate 'just cause' to defend liberal democracy against economic socialism and political communism. In contrast, now, it is the political left, usually represented by social-democratic parties (typically in coalition with the new green parties), who take the lead in jointly asserting a legitimate right to split society into 'new good vs. bad' camps. Namely, the 'new value claims' put forth by the new 'red-green' coalitions in most parliaments of liberal democracies today, and who mostly lead the international advocacy movement for climate-mitigating measures, are expressed in more or less explicit opposition against a new 'reactionary right-wing' political movement often held by those 'new liberals and progressives' to coincide even with a new 'populist and far-right' political wing, blamed for being potentially dangerous for contemporary societies and harmful for future generations.[32]

At the same time, the claim of the new political right that the above-states perceived climate change doctrine as mostly advocated for by the new political left would, in fact represent a (new) *totalitarian ideology* that may not be questioned,[33] reflects an almost symmetrical logic in comparison to the old Cold War politico-ideological antagonism. Implicitly criticising any sort of *ideological indoctrination*, such views herewith also serve as a warning of the *Orwellian nature* as experienced by the 'far left' in former communist-totalitarian regimes. What is also comparable to the old Cold War socio-political polarisation is that both camps blame one another for ideological biases and even deliberately lying, only this time it is about perceived 'scientific evidence' (or not) on climate change and global warming. Arguably, 'lying' always implies that one does not expose the whole truth but only the part of it that is in one's own camp's interest. In other terms, even under the terms of a perceived new Cold War, both sides remain caught in *'cognitive dissonance'* when representing their respective versions of climate change.

This is illustrated also in the *news media*. In the United States, it is generally held that the more left-wing camp, as for instance represented by CNN and the New York Times (NYT), generally resonate with the climate change 'believers', while FOX News and Breitbart are frequently associated with the climate change 'deniers'. What seems however symptomatic about the syndrome of *media polarisation* in the case of the USA, where the climate debate is increasingly characterised by 'emotional discourses',[34] is that it lacks political and ideological consistency, as both camps seem to represent their respective 'views & news' about climate change events and matters via a *selectivity bias*. This entails that also the *believers'* media outposts tend to cognitively filter what is not strictly within the range of their interests. For instance, on the occasion of the COP25 in Madrid (2019), CNN and the NYT did not report on the *climate change protests in Washington D.C.* on December 6, 2019, organised by climate NGOs such as 'Fonda Fire Drill' or even 'Shut Down D.C'.[35] In fact, this generic culture of selective news coverage by *both* 'believers' and 'deniers' may indicate that climate change is likely to enflame already highly polarising socio-economic and socio-political matters. One of these matters has to do with the socio-economic fallout of the Great Financial Crisis (GFC) unleashed at Wall Street in 2008, whose geoeconomic implications were felt particularly hard by the world's middle classes and the poor, and which also highlights systemic challenges to how the West rules the world.

Great Financial Crisis (GFC) and Western Worries to Preserve a 'Rules-Based International Order'

In October 2019, the Stanley Center for Peace and Security held a roundtable in Washington D.C. titled: 'The Next Global Financial Crisis and Climate Change: A Policy Agenda to Align with the Paris Agreement'. It examined the correlation between the financial system and the 'threat of climate change' and identified specific aspects of the regulatory and economic framework to provide for a (cit.) *'climate-safe world'*.[36] As indicated on top, especially in the Western literature, leading global economic and financial institutions, including the G7, G20, and IFIs, are now seen to play key roles in response to climate change. At the same time, as Tomlinson points out, success and failure will depend on whether these institutions remain capable of simultaneously *'defending a global, rules-based system'*.[37] Notably, as it is further underpinned, key financial and monetary institutions such as the IMF or even the Bank for International Settlements (BIS) *'must (!)* support strong adherence'[38] to this 'rules-based international order'. Since such an established order is *implicitly* meant to come under US leadership, these claims may, in turn, indicate some inflexibility among American elites to eventually share power and accept the new modalities of a *de facto* (though not yet formal) multipolar world, in which 'rules' may no longer be dictated unilaterally.

Beyond the USA, it indicates increasing *uncertainty* about Western global leadership, reflected also in financial and fiscal mismanagement and associated questions regarding the prospective role of the US dollar (USD) as the dominant reserve currency of the world economy. In fact, the lead Western powers and their preferred global governance institutions come under increasing pressure with shifting economic interdependence altering geopolitical parameters, a process unleashed with the GFC in 2008. With Martin Jacques, the GFC is crucial for the understanding of the relative decline of the USA and the comparative rise of China.[39] Thus, with 'the loci of decision-making shifting to other geographies... global policy-making and international cooperation are likely to be very different... given challenges to the existing global power structures'.[40] Consequently, geopolitical tensions, e.g., according to the above-mentioned Stanley Center, emerge with the *rise of China* and the corresponding *weakening of the EU post-BREXIT*, and with the *US*, e.g., since the Trump Administration, tending towards new *protectionism*, if not isolationism.[41] This means also that lead fora like the G7 and G20 are likely to 'fluctuate in importance... depending on the presidency at the time... (and since) ... emerging economies and developing countries will expect (e.g., from established Western leaders) to be involved in *global financial reform* of a *multipolar global economy*'.[42]

At the same time, and far from talking about *'de-growth'*, the global financial sector (under *de facto* Western leadership) has been preoccupied with the challenge to safeguard *'strong, sustainable, and balanced growth'*, e.g., via so-called 'supply of money' or what is otherwise termed *'Quantitative Easing' (QE)* to prevent deflation.[43] In this context, the BIS' *Basel III* financial reform package would recommend the *banking sector* to provide for *minimum liquidity requirements*, while monetary stability recommendations urged banks to focus on *low and stable inflation*.[44] However, large central banks, including the European Central Bank (ECB), bought significant quantities of *'carbon bonds'*, which in turn nurtured a political controversy at the European Parliament requesting the disclosure of the 'carbon footprint' of these assets, as Parliamentarians questioned 'why climate change risks were not taken into account when selecting bonds for purchase', [45] which appears more than a legitimate question, let alone from the standpoint of climate change 'believers'. Another controversy within the financial sector and concerning the new financial and environmental regulations post-GFC was if QE was the right way to go, or if rather, the main solution should be *'(Green) Credit Easing'*, e.g., for industry willing to invest in renewables and green technology. Meanwhile, from the perspective of *'sustainable finance'*, proposals are made for alternative QE, such as *'green assets' ('green QE')*, or even *'social assets' ('people QE')*. Finally, the deployment of *foreign exchange reserves* and *sovereign wealth funds*[46] are generally considered alternative solutions to QE and as appropriate drivers of a *green (industrial) transition,* along with *economic stimulus packages* and precautionary measures to rescue large banks during a future financial crisis ('too-big-to-fail' syndrome).

Others even question if *capitalist systems* themselves, let alone applying a 'market-fundamentalist' approach, can effectively set the right incentives to 'green the economy'. It appears as if the far-advanced *deregulation of financial markets*, characteristic in most Western societies today, generate inherent systemic constraints to stimulating green change. For instance, Jerneck has shown in a comparative study on the photovoltaic industries in the USA and Japan that in the case of the former, the so-called syndrome of *'financialization'* (when capitalist economies go largely 'unchecked') has hitherto tended to subordinate entrepreneurial innovation to financial speculation. This eventually resulted in the purchase of a substantial part of the American solar energy industry by competing industries abroad. In this context, and with reference to the economic theory of John Maynard Keynes, it is argued: 'If unchecked, innovative and profit-seeking financiers will break free of constraints and turn to *speculation* ... (in other terms) ... when the capital development of a country becomes the by-product of the activities of a *casino*'.[47] Ironically, most emerging economies are considered more apt to stimulate green economic growth, given that within their economic systems, the *state* retains substantial financial-regulatory competencies.[48] Accordingly, their comparative advantage in terms of *'economic statecraft'* is likely to place them in a prospective green growth leadership position.

New Geoeconomic Rivalries Lying at the Heart of 'Climate Geopolitics'

Reminding that 'Western economic dominance has been a recent historical phenomenon',[49] PWC characterises the *world's geoeconomic outlook (2015–2050)* as a *'global pivot'* to both the *East* and the *South*, with Asia and Africa showing the highest average economic growth rates.[50] Herewith, the PWC report, which implicitly mirrors the United States' *'Asia Pivot'* and corresponding *Global Rebalancing Strategy* adopted under the Obama Administration (Obama Doctrine[51]), stresses that the '*Asia-Pacific* ... (has indeed become) ... critical ... (and that with the) ... *rise of China from a regional to a global power projector*', this is (obviously) *challenging the traditional balance of power in Asia*,[52] let alone since the incoming new leadership at Beijing under Xi Jinping would launch the Belt- and Road Initiative (BRI) by 2013.

To say the least, climate-related policies turn out to be much more than a value-based blueprint for *'saving the planet'*. According to Bergamaschi & Sartori, the COP21 Paris Agreement is considered to even provide 'new and significant market opportunities'. Namely, so-called *'nationally determined contributions'* set at Paris represent a combined investment into renewables of at least $2 trillion per year over the next 15 years. At the same time, an *additional* $90 trillion in *infrastructure* investment will be needed by 2030 *'to meet global (sic!) growth expectations'*.[53] This also corresponds to the OECD report on *Financing Climate Futures: Rethinking Infrastructure* mentioning that the Paris-adjusted global infrastructure investment needs are herewith just about *10%* higher than if not adjusted ($6.9 trillion vs. $6.3 trillion per year, until 2030).[54] Against this background, Bergamaschi and Sartori further stress that especially the BRI represents a 'strategic initiative that places China at the heart of global infrastructure politics ... (while representing a) ... main vehicle of China's foreign policy interests'.[55] And since the existing global infrastructure is already responsible for some *60%* of global GHG emissions, '*greening the huge infrastructure pipeline* (of the BRI) is a *key challenge for global security* and prosperity ... (implying that) ... *major powers (are now likely to) battle to position themselves as China's leading green partner*'.[56]

Not only do prospective infrastructure investments by far outspend climate-related investments, and does the BRI represent the most important international infrastructure development project to date; China has also become the world's top *investor in green technology*.[57] Meanwhile, even the People's Bank of China (PBOC) emerges as a global leader in issuing *'green-asset-backed securities'*, considered to be a welcome 'pro-active measure',[58] as it tends to contrast with Western central banks' comparative green-financing regulatory hesitancy. Yet, the provision of what Western development institutions along with climate protagonists generally refer to as *'resilient infrastructure'* eventually goes hand in hand with a *'reset of the financial system'*, namely, *to* align financial flows with low-carbon targets.[59] However, since adjacent transformative ideas would be the 'rethinking of development finance for climate' along with the 'disentanglement of public budgets from fossil fuel revenues',[60] this could effectively coincide with a *Neo-Liberal Economic and Development Policy Agenda 2.0*. In this context, Sweeney denounces a *'green growth delusion'* (and implicit hypocrisy) among Western capitalist economies with regard to the records of the already established neo-liberal international financial and development institutions. Notably, *fossil fuel subsidies* have reached an impressive level of $5.2 trillion in 2017 alone.[61] Moreover, with the IMF prescribing *'efficient pricing'* (for energy markets), this is likely to produce 'zombie energy'.[62]

In other words, there is a tendency within the overall climate narratives, at least from a Western point of view, to essentially *fortify* the established world order, built since the 1980s on its typically *anti-government* and *anti-public* neo-liberal economic policies[63] and considering also the sometimes-devastating consequences of neo-liberal *Structural Adjustment Programmes (SAPs)* applied in developing and transitioning countries back in the 1980s and 1990s. Or, the emerging geopolitical discourse led by the West about the perceived need to deliver 'resilient infrastructure' coincided with *de facto 'Green Structural Adjustment'*, which from the perspective of the Global South is likely to read as *'Shock Therapy 2.0'*.[64] Nonetheless, as found by the International Institute for Sustainable Development (IISD), just *10–30%* of the global fossil fuel subsidies would suffice to 'pay for a global transition to clean energy'.[65] This is remarkable also in a broader global perspective, whereas the most advanced industrial economies of the Global North already consume 80% of the world's resources, which means just by about 20% of the world population.[66]

The underlying geopolitical and geoeconomic rivalries must also be considered in view of *energy markets*, and especially the development of *energy prices*. Just about the time when Alan Greenspan was to retire as chair of the FED in 2006, he expressed his concern about a 'growing instability of global energy prizes'.[67] In fact, as a direct result of the Anglo-American invasion of Iraq in 2003,[68] by around 2008–2011, energy prices were almost six times higher than if compared to certain levels during the 1990s. While this considerable price spike in energy markets timely coincided with the outbreak of the GFC, geopolitically, it occurred during a time characterised by the comparative *'strength and assertiveness of China'* in contrast to *'American economic and military fatigue'* and the *'disarray of Europe'*, (e.g., with the spread of the financial crisis over the Atlantic to the Euro-zone), thus 'pushing the West into a New Cold War'. Consequently, as China became the new focus of US geostrategy, the new global energy market dynamics required the US to pay attention to Saudi Arabia (oil) and Russia (natural gas[69]) from a geoeconomic point of view, as America became worried about the 'future of US gas flows'.[70] As a result, geopolitical rivalries over oil and natural gas markets are likely to remain a predominant feature of international energy politics, regardless of how much both established and emerging economies are investing in renewables. According to IEA's projections, by 2035, fossil fuel use will still account for about *three-quarters* of global energy use, with a total increase of about *one-quarter* of global fossil fuels by then.[71] After all, the eventual collapse of public opinion regarding climate change in the USA post the financial crisis by 2009 also coincided with the failure of the incoming Obama Administration to take the lead at the Durban climate conference, where US negotiators showed 'little enthusiasm' regarding climate change accordingly.[72]

Meanwhile, a study carried out in the UK by the Financial Stability Board and its Prudential Regulation Authority shows that only about 10% of banks are taking a 'strategic' and forward-looking stance on climate risks. This, in turn, corresponds with the ECB reporting that 'the overwhelming majority of banks have not yet established internal procedures that allow them to systematically identify and manage climate-related risks'.[73] Put in an even a different perspective, the US FED FSB's Task Force on Climate-Related Financial Disclosures (TCFD) finds that the 500 largest companies by market capitalisation seem to be exposed to (just) $1 trillion in risk over a 5-years-period.[74] This would equal some $250 billion per year for those top 500, which seems to be peanuts compared to their combined revenues, considering also that just during the global coronavirus pandemic alone, the world's billionaires increased their combined wealth by almost $40 trillion according to the South Centre. After all, with reference to the UNEP, the damage to the global economy since 2011, would be around $28 trillion per year if the established scale of unsustainable economic activity would be maintained.[75] Thus, when someone claims that climate change causes 'damage to the economy', one is well advised to have a closer look and eventually differentiate which of the economy's unit levels is concerned.

Western-lead Climate Advocacy Won't Suffice to Reclaim Eurasia's Heartland

Throughout the past two centuries, the Anglo-American geostrategy of Britain and the USA has focused on the control of Eurasia (and therein, containing its main continental power, Russia) and especially the mineral-rich Eurasia's heartland,[76] a central geographic area in continental Eurasia encompassing Russia's Central Siberian plains and coinciding with the so-called *'strategic ellipse'* where over half of all fossil fuel reserves of the world are concentrated. This extends to Central Asia, the Caspian Basin, and the Persian Gulf region. Against the background of great power rivalries and according to contemporary geopolitical parameters, Europe will end up being outmanoeuvred, given its ambitious targets to *fully exit* from fossil fuels. Since no reductions really imposable on Russia, nor the USA, China, and India – the world's largest GHG emitters – whatever the EU decides now, it won't really 'carry way' to alter climatic patterns.[77] And in the event of the two largest territorial countries on earth, the Russian Federation and Canada, they are likely to even become *net 'winners'* of climate change and global warming. For instance, Russia, who according to Bosnjakovic, 'increasingly thinks of itself as an Asian nation',[78] will benefit from extending its agricultural yields and take the opportunity to enhance and modernise agricultural and food production and exports, as well as expand its exploration and exploitation of fossil fuel energy resources[79] – and since especially Asia and Africa (the two fastest growing regional economies and only net-growing demographics) will badly need to safeguard both *energy and food security*. Herewith, Eurasia's geopolitical chessboard is experiencing a *continental pivot (within Eurasia) to the East* (along with a *maritime power shift from the North-Atlantic to the Indo-Pacific*). In this context, the *Shanghai Cooperation Organization (SCO)*, jointly lead by China and Russia, along with the new *Asian Infrastructure Investment Bank (AIIB)* launched under Xi Jinping in 2015 and co-led by China and India, demarcate the dominant institutional pillars of Eurasia's new geopolitical puzzle. If during the 20th century, the geopolitical patterns of the largest and most central continent of world political geography would read *EURasia*, during the 21st century, its geopolitical and geoeconomic chessboards will increasingly read *eurASIA*.

In fact, Eurasia is entering a new era in which Europe does not even dominate its immediate neighbourhood of the *Mediterranean* anymore. In this context, as pointed out by Schaffran, the Euro-Med Contact group would have to be widened to accommodate old and new influential actors, including not only the USA and a re-assertive Russia, but also China, as well as Qatar, Iran, and Saudi Arabia.[80] Furthermore, the recent diplomatic rapprochement between Iran and Saudi Arabia facilitated by China suggests the emergence of a new order in the *Middle East*,[81]though this time framed from Eurasia's East (Asia) and no longer from Eurasia's West (Europe, North-Atlantic). Finally, with *Iran* acquiring full membership at the SCO this year,[82] an entirely new geopolitical and geoeconomic Eurasian chessboard is in the making that resonates deep into South Asia and the Indian Ocean Area (IOA). While the SCO further extends its already impressive range to one of the key energy players in *West Asia* and the *Persian Gulf*, back in 2017, the SCO had already expanded to *South Asia*, when India and Pakistan joined, rendering the SCO by far the largest regional organisation in the world by territorial, demographic, and economic standards, encompassing 40% of the world population and a quarter of global GDP. As Ahmed et al. point out, including with comparative perspectives on the South Asian Association for Regional Cooperation (SAARC), which notoriously suffers from regional disagreement about leadership along with the persistent strategic arms race between India and Pakistan, in contrast, and given that both China and Russia keep strategically investing in the SCO to reinforce their comprehensive strategic partnership along with their economic interests in Eurasia, this new constellation in Asia is likely to be *'challenging the dominant Western world order'*.[83]

For *India*, a key emerging power in Asia, membership in the SCO allows Delhi to 'build strategic regional partnerships', while extending its already existing membership in new strategic groups of the Global South such as the BRICS and the India-Brazil-South Africa (IBSA) Dialogue. Via the SCO, India may underpin its involvement in Afghanistan's reconstruction process and serve as a platform to reach out to Iran, including via the strategic port of Chabahar, which connects the Indian Ocean with the North-South Transport Corridor ('from St. Petersburg to Mumbai'). Iran's new membership in the SCO offers *'crucial opportunities for India'*.[84] SCO membership also represents an opportunity to more effectively balance against China, its main regional rival in Asia, especially since especially the China–Pakistan cooperation along the BRI, with the China–Pakistan Economic Corridor (CPEC), is considered by Delhi as part of a potential strategic encirclement by China, along with third BRI economic corridors to the Indian Ocean established in Bangladesh and Myanmar.

Notably, the Indian economy keeps growing considerably, with over 7% of GDP per year, currently outpacing the Chinese economy. According to the WEF, India had already by-passed Canada, Brazil, and Italy in 2017 in terms of nominal GDP, and by 2030, projected to further surpass the United Kingdom, France, Germany, and even Japan[85] and herewith become no. 3 behind China and the USA,[86] the country's main priority is to (cit.) *'secure the necessary energy resources'*.[87] And its newly gained membership in the SCO will essentially facilitate India's energy cooperation with both Central Asia and Russia[88]: 'Access to the resource-rich Central Asian region is a key component of India's quest for *energy security*, a quest that has drawn India towards the SCO'.[89]

Given India's geoeconomic rivalry with China in Asia, Delhi concluded a *Civilian Nuclear Agreement* with the United States by 2009 and more recently also agreed to a *Logistics Exchange Agreement* with the Pentagon in 2017,[90] two strategic cooperation initiatives that may further weigh on its already laden regional rivalry with Pakistan. At the same time, the flexible multilateral format of the SCO allows both India and Pakistan to provide for more confidential diplomatic talks and *security-building measures*, including in their sensitive and joint border region with China, and both countries may use the SCO platform for their mutual energy outreach to Central Asia. In fact, after China had been initially opposed to Indian membership in the SCO, it later compromised, not only to bring Pakistan into the bloc, but since Beijing's new strategic vision under Xi Jinping eventually envisages a potential '*alliance* with India' (and since) 'improving India–China relations (according to the vision of Xi Jinping) would be a *historical mission*'.[91]

This *de facto*, though not (yet) formal, India–China strategic coordination is also expressed in India's role as *no. 2 at the AIIB*. Along with the triangular strategic ties of both countries with Russia and Iran, a new Eurasian 'continental giant' is in the making that could be compared to the conventional Western *trilateral* format of *US–EU–Japan*,[92] which ever since the Cold War has become the geostrategic centrepiece of Western global strategic coordination, institutionally represented also by the New York-based *Trilateral Commission* co-founded by David Rockefeller and Zbigniew Brzezinski. Meanwhile, the combined GDP of the central strategic group of the Global South, the BRICS, has caught up to that of the G7. And despite the border standoff at Doklam in 2017, Delhi has kept up with diplomatic engagement with Beijing, shown for instance by India's attendance at the BRICS Summit in Xiamen the same year – in the context of which China notably co-sign a Joint Declaration against terrorism condemning three Pakistan-based terrorist organisations.[93]

Finally, joint membership in the *'energy club'* of the SCO has also proven beneficial for improving even Indian–Pakistani relations themselves, since the two countries do have 'shared interests in gaining *energy connectivity*' with Central Asian countries, including Tajikistan and Kazakhstan. In this context, it is notable to mention the joint attendance of India and Pakistan at the inauguration of the *Tajikistan–Afghanistan–Pakistan–India (TAPI) natural gas pipeline* in February 2018. Also, one could refer to the project of the *Iran–Pakistan–India (IPI) natural gas pipeline*, which now could gain new momentum with Iran's effective integration into the SCO. Beyond energy security, the SCO also serves India for *economic integration with Eurasia*, e.g., via a prospective free-trade agreement with the Moscow-led *Eurasian Economic Union (EEU)*.[94]

Conclusion: Great Powers Keep Focusing on Energy Security: Lessons from the Indo-Pacific

Considered through the lens of the new global balance of power dynamics with the increasing epicentre of the Indo-Pacific, geopolitical and geoeconomic realities on the ground suggest that climate mitigation efforts aren't really a priority for the great powers to date, neither for emerging economies nor for established industrial economies. Far from effectively transitioning towards a tangible low-carbon green economy for the foreseeable future, great powers continue engaging in strategic rivalries over energy market and supply chain control, with special emphasis on key fossil fuels, especially oil and gas. From a geostrategic point of view, this *de facto* continued prioritisation of 'energy security' over 'climate security' is intrinsically related to a global pivot of energy markets to Asia, especially concerning fossil fuel exports from the Gulf region to increasingly Asia-centred world markets. At the same time, as Asia's fossil fuel demands continue to increase compared to other world regions, and as China has become the largest oil importer in the world, this hasn't prevented emerging economies from allocating strategic investments also into renewables and green technology and thus diversifying their energy spectrum and portfolio. However, distinct from Europe, such measures must be considered more as a *diversification* of the energy portfolio alongside an investment in strategic technologies, rather than a plan to suspend fossil fuel use altogether. At the same time, in terms of global governance and multilateral cooperation,[95] China is increasingly replacing the USA in effectively leading worldwide *climate awareness*.

Regarding the Indo-Pacific, China has been reaching out to the Indian Ocean via the BRI mainly to circumvent a potential *Malacca Dilemma*[96] with the USA in the future and to safeguard access to the Gulf region's fossil fuels, especially oil – since over 80% of its conventional oil imports have been shipped through the Strait of Malacca. Meanwhile, India, with its unique demographic and economic growth rates, is deploying its membership at the SCO to establish new *energy diplomacy*' to gain access to Central Asia's rich fossil fuel reserves. India is increasingly cooperating with both Iran and Russia, including in the delivery of the NSTC. With *Asia's two giants*, which are by far the largest demographics in the world and rejuvenating to great power status, conventional Western powers, including the USA and Australia, keep looking to the IOC predominantly through a geopolitical and geoeconomic lens. While Western power blocs in Europe, including NATO and the EU, now tend to see *China* as a prime geopolitical rival, the West is attempting to *engage India* in a potential 'alliance of democracies' to contain perceived 'authoritarian regimes' in Eurasia, including Russia and China. However, Russia's re-assertiveness and China's rise are just the beginning of what others call the *'rise of the rest'* (of the Global South),[97] including the emergence of India as a new great power in Asia. Not even the main European economies, including France, Germany, and the United Kingdom, can afford to put 'climate protection' over these new geopolitical patterns, in the context of which control over energy markets, connectivity infrastructure, and technological supply chains remain central to geostrategy. At the same time, a new bipolarity and superpower competition is eventually crystallising, in the context of which China has been catching up with the USA in terms of *Big Tech*, including in the fields of Information and Communication Technologies (ICT), as well as *cyber sovereignty*.[98]

In conclusion, for the *USA*, which hitherto occupied a hegemonic role in the IOR since the collapse of the Soviet Union, the main objective is to *contain China* from becoming the new dominant power in this region while also keeping America the dominant power in Asia overall and beyond. This at least seems to be the prime geostrategic goal according to certain neo-conservative national security advisors at the White House and their fellow defence planners at the Pentagon, whose vision since the turn of the 21st century has been to provide for a *New American Century* – despite the fact that such a plan may not seem to benefit from broad public support in the USA.[99] In any event, to effectively rival China in the Indo-Pacific and Asia overall, the US, increasingly sidelined by China at rank no. 2 in geoeconomic terms, will have to foster its strategic partnership with India. However, India's sustained collaboration remains unpredictable for Washington D.C., given India's long-standing leadership within the *non-aligned movement* and its newly found economic prowess and *regional leadership ambitions*. Beyond, America's declared normative ambition is to take leadership in defence of what is now generally referred to as a 'rules-based international order',[100]against the background of which the global West seeks to expand NATO not only to Eastern Europe but also to the Arctic, Africa, and even the Indo-Pacific. At present, Biden's new *Indo-Pacific Economic Framework* (IPEF) seems to more represent a paper tiger than a convincing alternative to China, which tends to win the geo-commercial race against the US in the Indo-Pacific. Anyway, since the USA ultimately attempts to create a 'global alliance of democracies' with the potential to even push aside the established multilateral framework of the United Nations,[101] this in turn reflects a 'Neo-Cold War' game against a perceived *new Eastern bloc*, at the core of which the ever-deepening *Chinese-Russian comprehensive strategic partnership* in Eurasia stands out. This partnership institutionally coincides with the impressive Shanghai Cooperation Organization (SCO) led by Beijing and Moscow. For the Indo-Pacific, this means that the USA will increasingly have to convince key regional allies from the Asia-Pacific, especially Japan and Australia, to strategically engage with the IOR, while in turn advocating for India's strategic engagement with the South China Sea (SCS).

For *China*, its strategic outreach via the BRI and corresponding economic corridors to the IO has less been nurtured by military calculations, such as strategically encircling India via a perceived 'String of Pearls'[102] and to proliferate commercial–military dual-use ports across the IO and India's vital neighbourhood accordingly (Pakistan, Bangladesh, Myanmar, and Sri Lanka), than by a geoeconomic rationale of preserving strategic commercial routes leading through the IO in the context of an escalating *'trade war'* with the USA. In this context, the BRI itself has been a direct response under the new Chinese leadership of Xi Jinping in 2013 to the USA's 'pivot to Asia' and geostrategic rebalancing from the North-Atlantic to the Asia-Pacific under the Obama Administration.[103] With the BRI's grand geoeconomic design and given its unparalleled connectivity and logistics dimensions (going even far beyond the range of the US' Marshall Plan in Europe after WWII), one of the main objectives for the Middle Kingdom has been to avoid direct (military) confrontation with the USA in the SCS and to maintain focus on economic development. Not only does China have BRI-related cooperation agreements with all member states of the *Gulf Cooperation Council (GCC)*; its recent success in brokering some rehabilitation of diplomatic relations between Iran and Saudi Arabia can also be considered a major step towards stability and peace in the Persian Gulf Region and the Middle East overall, as it may provide new impetus also to the OPEC+ format. China's increasingly dominant geoeconomic assertiveness in Asia and beyond is also reflected in its successful participation in the *Regional Comprehensive Economic Partnership (RCEP)* initiative under the format of the group of ASEAN+. In fact, the success of the RCEP contrasts with the failure of the TPP launched under Obama and suspended under Trump, as it *ex ante* precluded China. Interestingly, since Tokyo has recently taken leadership to eventually relaunch the TPP under the new *Comprehensive and Dynamic TPP* (CDTPP), keeping doors open to Beijing for future accession, this clearly shows the new multipolar power dynamics in the Asia-Pacific, in which the USA remains at best a *primus inter pares*.

Thus, Western powers tend to keep a neo-Cold War, and therefore more confrontational, *chess-game*-like course on China and other geopolitical rivals; however, they end up frustrated this time, since the contemporary Middle Kingdom, strategically unified with Russia, happens to be another 'caliber' if compared to the former Soviet Union, a new bloc that the global West can't effectively contain this time. Meanwhile, strategic planners in

Beijing have preferred to avoid a neo-ideological fight with the West, and therefore play a less confrontative *Go-game*-type game, especially with American counterparts, deploying a grand geoeconomic strategy backed by economic statecraft. The result of this asymmetrical double-game is likely to write the textbooks of a new world order in the making with potentially Sino-Asian characteristics. And if Julius Caesar was right, as the global West (15% of the world population) may wish to 'believe in climate change' (though mainly for geopolitical reasons), the priority of the Global South (85% of the world population) remains to believe in its legitimate 'right to development'.

Notes

1 Word Wildlife Fund (WWF).
2 World Health Organization (WHO).
3 The Lancet.
4 Patraeus & Benedict, 2021, p. 3–4. Emphasis added by author.
5 Patraeus & Benedict, 2021, p. 4.
6 PriceWaterHousCoopers (PWC), 2016. Emphasis added by author.
7 Tomlinson, 2019, p. 1.
8 Halden, 2007, p. 56.
9 Halden, 2007, p. 9.
10 See also the author's comment on the IPCC in the next section.
11 Halden, 2007, p. 150.
12 Halden, 2007, p. 151.
13 Notably, there is an implicit distinction made between geographic regions and strategic groups of countries.
14 Halden, 2007, p. 6.
15 Considering that the populations of China and India are almost three times larger than the combined demographics of the US, EU, and Japan, it is important to account for the substantial market potential and influence these countries possess.
16 These countries have been crucial, for instance to upholding the 'Petro-dollar' world economic system. Notably, GCC member states emit the highest average *per capita* GHG emissions in international comparison, as no fewer than six among them (Qatar, Kuwait, UAE, Bahrain, Saudi Arabia, and Oman) figure among the top ten; Qatar is ranking by far at position No. 1, followed by Kuwait (No. 2), the UAE (No. 3), and Bahrain (No. 4). The remaining countries among the world's top ten per capita emitters are all Western countries: Canada (No. 5), Australia (No. 6), Luxembourg (No. 7), and the USA (No. 9).
17 Halden, 2007, p. 6.
18 Bonner, 2014. Emphasis added by author.
19 US Government, Patriot Act.
20 Busch & Judick, 2021, p. 15.
21 Center for Climate and Energy Solutions, 'Realities vs. Misconceptions about Climate Change'.
22 Gavrilo et al., 2016, p. 307.
23 Gavrilo et al., 2016, p. 306. Emphasis added by author.
24 Busch & Judick, 2021, p. 9. Although while fracking methods can hardly be 'better for the environment' overall.
25 Busch & Judick, 2021, p. 11. For instance, refugees from the MENA region are severely affected by protracted drought. This argument, however, implies that 3rd

regions, including Europe and parts of Asia, may be disproportionately exposed to climate migrants.

26 Busch & Judick, 2021, pp. 11–19.
27 Centeno, 2020, p. 4.
28 Centeno also finds that enhanced participation of women in climate negotiations hasn't really altered the way climate change is addressed.
29 Centeno, 2020, p. 4. Emphasis added by author.
30 Centeno, 2020, p. 13.
31 Maechler & Moser, 2019, p. 3.
32 Yet, what is portrayed as the 'new far right' in this context can hardly be compared to the political one-party totalitarian regimes historically experienced, for instance in fascist Italy or even Nazi Germany.
33 Busch & Judick, 2021, p. 14.
34 Esparcia et al., 2021, p. 4.
35 Esparcia et al., 2021, p. 9.
36 Stanley Center for Peace and Security, 2019. Emphasis added by author.
37 Tomlinson, 2019, p. 9.
38 Tomlinson, 2019, emphasis added by the author. This normative narrative, from a Western point of view, may indicate that established global financial and monetary institutions may increasingly come under pressure from the emerging economies of the Global South.
39 Jacques, 2009.
40 Jacques, 2009.
41 Stanley Center, 2019.
42 Tomlinson, 2019, p. 9. Emphasis added by author.
43 Kern & Fisher, 2020, p. 1.
44 Kern & Fisher, 2020.
45 Kern & Fisher, 2020, p. 8.
46 Kern & Fisher, 2020.
47 Jerneck, 2017, p. 2. Emphasis added by author.
48 Campiglio, 2014, p. 1.
49 PWC, 2016, p. 5.
50 PWC, 2016, pp. 3–5.
51 This (yet unofficial) 'Obama Doctrine' essentially posits that China has effectively become America's prime international 'enemy image' since the Obama Administration.
52 PWC, p. 6. Emphasis added by the author. Notably, with the established system of balance of power in Asia, the USA, thus an overseas power, has been the hegemonic power since the collapse of the Soviet Union in 1991. This situation also corresponds to Europe's geopolitical chessboard ever since WWII, and in the context of which the US more recently has been preoccupied with a re-assertive Russia under President Putin.
53 Bergamaschi & Sartori, 2018, p. 10. Emphasis added by author.
54 OECD, 2018, p. 2.
55 OECD, 2018, p. 10.
56 OECD, 2018, Emphasis added by author.
57 Hilty & Bieser, 2017, p. 5.
58 Kern & Fisher, 2020, p. 6
59 Regarding the broader context of financial transition, especially since the global coronavirus pandemic, see also the 'Great Reset' suggested by the World Economic Forum, or even the IMF's 'Great Transformation'. Notably, the joint vision of the Great Reset and the Great Transformation is essentially about a consequent digitalisation of the economy (WEF; IMF).

60 OECD, 2018.
61 Sweeney, 2020, p. 87.
62 Sweeney, 2020, p. 88.
63 Sweeney, 2020, pp. 88–89.
64 Sweeney, 2020, p. 90.
65 Sweeney, 2020, pp. 87–88.
66 Headrick, 2019, p. 47.
67 Jones et al., 2014, p. 5.
68 Jones et al., 2014, p. 11.
69 From a geopolitical perspective, one could also add Iran in this context, which, next to Russia and Qatar, has one of the largest natural gas reserves in the world.
70 Jones et al., 2014, p. 9.
71 Jones et al., 2014, p. 6.
72 Bosnjakovic, 2012, pp. 8–9.
73 FSB, 2020, p. 30.
74 Brainard, 2019, pp. 5–6.
75 Kern & Fisher, 2020, p. 2.
76 The so-called 'heartland theory' was first formulated by the British world geographer, Halford Mackinder, by the turn of the 20th century. By this time, notably, Britain's main geopolitical rival in continental Europe wasn't so much Russia, but rather Germany.
77 Bosnjakovic, 2012, p. 23.
78 Bosnjakovic, 2012, p. 10.
79 Bosnjakovic, 2012, p. 9.
80 Scheffran, 2020, p. 61.
81 Fantappie & Nasr, 2023.
82 Global Times, 2022.
83 Ahmed et al., 2019, p. 5.
84 Ahmed et al., 2019, p. 10.
85 Ahmed et al., 2019, p. 11.
86 According to Standard Chartered, India's GDP is even likely to be larger than that of the USA by 2030. Standard Chartered.
87 Ahmed et. al., 2019, p. 11.
88 Ahmed et al., 2019.
89 Ahmed et al., 2019, p. 10. Emphasis added by author.
90 Ahmed et al., 2019.
91 Ahmed et al., 2019, p. 12. Emphasis added.
92 Which is also the core of the Organisation for Economic Cooperation and Development (OECD), the most advanced industrial economies.
93 Ahmed et al 2019, , pp. 12–13.
94 Ahmed et. al., 2019, pp. 13–14.
95 For instance, Chinese President Xi provided a historical key address to the WEF in Davos 2017, assuring the world of firm Chinese support of multilateral diplomacy, at a time when US President Trump proclaimed 'America First'. Mathros.
96 Ahmed et al., 12.
97 Zakaria, 2008.
98 This may also reflect the formation of a 'G2' between China and the USA.
99 Barry, 2006.
100 Notably, as the very term 'rules' suggests, it is not about a *formal-legal* framework, e.g., as based on the UN Charter and international law more generally.
101 In fact, both neo-conservative and *de facto* neo-liberal factions in US elites, including from both parties, increasingly express their more or less explicit

dissatisfaction with the UN system and multilateral diplomacy. Unlike the situation after WWII, today, the US can't maintain a clearly dominant leadership position there, given that both Japan and the EU have become major donors to development aid and humanitarian relief, and given emerging economic powers' increasing assertiveness in multilateral negotiations. This is illustrated by the Doha round after China's full membership in the WTO in 2001, despite the fact that America nominally contributes the largest financial part to the budget of most established international organisations.

102 In fact, the concept of 'String of Pearls' was originally developed in the Pentagon as part of the US armed forces' global military assets, including naval bases, and to encircle its main geostrategic competitors in Eurasia, including China, Iran, and Russia.

103 Ahmed & Lambert, 2022.

References

Ahmed, F., & Lambert, A. (2022). *The belt and road initiative geopolitical and geoeconomic aspects*. Routledge.

Ahmed, Z. S., et al. (2019). Conflict and cooperation? India and Pakistan in Shanghai cooperation organisation. *Pacific Focus, XXXIV*(1), 5–30.

Barry, T. (2006, June 28). *Rise and demise of the new American century*. University of Alberta, IRC Special Report. https://sites.ualberta.ca/~raitken/documents/0606riseanddemise.pdf.

Bergamaschi, L., & Sartori, N. (2018). *The geopolitics of climate: a transatlantic dialogue* (IAI Papers 18, no. 10). Istituto Affari Internazionali.

Bonner, B. (2014, August 8). A lesson from Caesar: Investors will believe what they want. *Money Week*. https://moneyweek.com/332975/bill-bonner-a-lesson-from-caesar.

Bosnjakovic, B. (2012). Geopolitics of climate change: A review. *Thermal Science, 16*(3), 629.

Brainard, L. (2019, November 8). *Why climate change matters for monetary policy and financial stability*. Board of Governors of the Federal Reserve Systen, San Francisco, CA, Speech. https://www.federalreserve.gov/newsevents/speech/brainard20191108a.htm

Busch, T., & Judick, L. (2021). Climate change – That is not real! A comparative analysis of climate-sceptic think thanks in the USA and Germany. *Climate Change, 164*, 18. https://doi.org/10.1007/s10584-021-02962-z

Campiglio, E. (2014, June). *Beyond carbon pricing: The role of banking and monetary policy in financing the transition to a low-carbon economy* [Working Paper No. 181]. Centre for Climate Change Economics and Policy.

Centeno, E. (2020). The socio-political construction of climate change: Looking for paths to sustainability and gender justice. *Sustainability, 12*, 3382. https://doi.org/10.3390/su12083382.

Center for Climate and Energy Solutions. (2012). *Realities vs. misconceptions about climate change*. https://www.c2es.org/wp-content/uploads/2017/03/misconceptions-realities-climate-science-06-2012.pdf

Economics Help. (2021). *Top CO2 polluters and highest per capita*. https://www.economicshelp.org/blog/10296/economics/top-co2-polluters-highest-per-capita.

Esparcia, A., et al. (2021). Public opinion about climate change in United States, partisan view and media coverage of the 2019 United Nations Climate Change Conference (COP 25) in Madrid. *Sustainability, 13*, 3926. https://doi.org/10.3390/su13073926

Fantappie, M., & Nasr, V. (2023, March 22). A new order in the Middle East? Iran and Saudi Arabia's rapprochement could transform the region. *Foreign Affairs*. https://www.foreignaffairs.com/china/iran-saudi-arabia-middle-east-relations.

Financial Stability Board. (2020, November 23). The implications of climate change for financial stability. https://www.fsb.org/2020/11/the-implications-of-climate-change-for-financial-stability

Gavrilo, M., et al. (2016, March). *Global warming – Between the myth and reality*. ResearchGate, Conference Paper. https://www.researchgate.net/profile/Milivoj-Gavrilov/publication/308986389_GLOBAL_WARMING_-_BETWEEN_THE_MYTH_AND_REALITY/links/57fcdbbc08aeb857afa09342/GLOBAL-WARMING-BETWEEN-THE-MYTH-AND-REALITY.pdf

Global Times. (2022, July 12). Iran to obtain full SCO membership in 2023: Embassy – Concerted efforts to reject "hegemony and unilateralism" win hearts., https://www.globaltimes.cn/page/202207/1270360.shtml

Halden, P. (2007, December). *The geopolitics of climate change* (FOI-R-2377-SE). Swedish Defense Research Agency (FOI). https://www.klimatilpasning.dk/media/1154179/The%20geopolitics%20of%20CC.pdf

Headrick, D. (2019). Climate change: Debate and reality: Challenges to the international system. *International Review of Environmental History, 5*(1), 43–60.

Hilty, L., & Bieser, J. (2017, October). *Opportunities and risks of digitalization for climate protection in Switzerland*. University of Zurich, Informatics and Sustainability Research Group.

International Monetary Fund (IMF). (n.d.). *From the great lock-down to the great transformation*. https://www.imf.org/en/News/Articles/2020/06/09/sp060920-from-great-lockdown-to-great-transformation

Jacques, M. (2009). *When China rules the world: The rise of the Middle Kingdom and the end of the western world*. Penguin Books.

Jerneck, M. (2017). Financialization impedes climate change mitigation: Evidence from the early American solar industry. *Science Advances, 3*, e1601861.

Jones, B., et al. (2014, March). *Fueling a new order? The new geopolitics of security consequences of energy*. Project on International Order and Strategy.

Kern, A., & Fisher, P. (2020). Central banking and climate change, chapter 3. In P. Pisher (Ed.), *Making the financial system sustainable* (pp. 49–74). Cambridge University Press. https://doi.org/10.1017/9781108908269.006

Maechler, A., & Moser, T. (2019, November 14). *Climate risks and central banks: An SNB perspective*. Swiss National Bank (SNB).

OECD, UNEP, & World Bank Group. (2018). *Financing climate futures: Rethinking infrastructure, policy highlights*. https://www.oecd.org/environment/financing-climate-futures-9789264308114-en.htm

Patraeus, D., & McAleenan, B. (2021). *Climate change as a growing force in geopolitics*. Policy Exchange. https://policyexchange.org.uk/wp-content/uploads/2022/10/Environmental-Affairs-the-Geopolitcs-of-Climate-Change.pdf.

PriceWaterHousCoopers (PWC). (2016, November). *Five megatrends: And their implications for global defense & security*. www.pwc.co.uk/megatrends

Scheffran, J. (2020). The geopolitical impact of climate change in the Mediterranean region: Climate change as a trigger of conflict and migration. In *Climate change in the Mediterranean, yearbook 2020* (pp. 55–61).

Standard Chartered. (2023, January 12). Lack of climate adaptation investment could cost emerging markets hundreds of billions by 2030. https://www.sc.com/en/media/press-release/lack-of-climate-adaptation-investment-could-cost-emerging-markets-hundreds-of-billions-by-2030

Stanley Center for Peace and Security. (2019,). *Global financial crisis and climate change: A playbook for action* 60th Strategy for Peace Conference, Warrenton VA, USA, October 16-18, 2019.
Sweeney, S. (2020). Weoponizing the numbers: The hidden agenda behind fossil-fuel subsidy reform. *New Labor Forum*, 29(I), 87–92.
The Lancet. (n.d.). *Planetary .health* https://www.thelancet.com/journals/lanplh/home
Tomlinson, S. (2019, September 20). *The geopolitics of climate change*. https://www.e3g.org/publications/the-geopolitics-of-climate-change-unsg-climate-action-summit
World Economic Forum (WEF). (n.d.). *Great reset*. https://www.weforum.org/great-reset
World Health Organization (WHO). (n.d.). *One health*. https://www.who.int/news-room/questions-and-answers/item/one-health
World Wildlife Fund (WWF). (n.d.). *Focus: Biodiversity*. https://wwf.panda.org/discover/our_focus/biodiversity/biodiversity
Zakaria, F. (2008). The rise of the rest: the financial crisis aside, China, India, Brazil, and much of the developing world is booming. It seems scary, but writer Fareed Zakaria thinks it's actually good news for the U.S. *Gale Academic Online*. https://go.gale.com/ps/i.do?id=GALE%7CA191098336&sid=googleScholar&v=2.1&it=r&linkaccess=abs&issn=15251292&p=AONE&sw=w&userGroupName=anon%7E7d8b3c8d.

3 Climate Change in APEC

Impacts and Adaptation Strategies

Pittaya Suvakunta

Introduction

According to the Asia-Pacific Economic Cooperation (APEC), a regional economic conference, member economies in the Pacific Rim must collaborate to achieve sustainable development. The APEC region's various businesses, landscapes, and communities are all under the threat of climate change, which is a significant concern. APEC focusses on climate change, sustainable development, and environmental preservation as significant discussion topics.[1] Furthermore, APEC recognises the importance of striking a balance between environmental concerns and economic advancement. Climate change can have severe consequences, including extreme weather, depletion of natural resources, livelihood loss, and food insecurity, with developing economies being the most negatively impacted. Moreover, climate change is believed to be a contributing factor to the growing frequency of natural disasters, which account for approximately 70 per cent of the Asia-Pacific region. APEC economies consume over 60 per cent of the world's energy, so they must explore ways to reduce their reliance on fossil fuels to meet their energy needs. As the region gets more industrialised and more people migrate from rural to urban areas, these consumption levels are anticipated to rise.

The Asia-Pacific area is being severely impacted by rising sea levels, excessively high temperatures, and extreme weather events connected to climate change. These factors damage the region's economies, physical assets, and natural resources, worsening developmental issues like poverty, food and energy security, and health. The Asia-Pacific area is home to several nations that are both highly sensitive to the negative effects of climate change and geographically fragile. Climate change might push over 100 million people into extreme poverty by 2030 if climate change is not addressed via climate-oriented development, negating the progress made in the previous decades to reduce poverty.[2]

The Asia-Pacific area is particularly vulnerable to the effects of risks due to increased exposure of individuals and their belongings. Despite decades of economic expansion, persistently high socio-economic vulnerability, poor adaptive capability, and low incomes further exacerbate the situation. In emerging nations, unplanned urbanisation-related expansion of hazardous

DOI: 10.4324/9781003586869-4

regions has been the main factor contributing to greater exposure, a significant problem in Asia-Pacific. The current 40-45 per cent urbanisation rate is predicted to rise to 50 per cent in the next 10 years.

The essence of international security has undergone significant changes in the past 50 years. Conventional military conflicts between countries are no longer central to security discourse. The incidence of intra-state conflicts has increased since the end of World War II. This trend can be attributed to decreased funding and support for state-to-state proxy warfare after the end of the Cold War. Natural resources have contributed significantly to the dynamics of many internal and international wars that occur nowadays, with most of these conflicts being self-financed. For instance, a United Nations Environment Programme (UNEP) research claimed that 18 conflicts had broken out since the end of the Cold War, with 40 per cent of those conflicts occurring over the last 60 years due to issues with natural resources.

The 1982 UN General Assembly World Charter for Nature recognised the potential negative relationship between natural resources and security and stated that conflict results from the struggle for scarce resources. More than 20 years later, former UN Secretary-General Kofi Annan established the High-Level Panel on Risks, Challenges, and Change to analyse current threats to peace and security. The panel's 2004 report, A More Secure World, Our Shared Responsibility, summarised its findings and stated that high population concentrations and resource constraints, particularly concerning land and other natural resources, can exacerbate environmental stress and civil unrest.

Although there is general agreement that the environment and security are related, this relationship is mainly ignored in the present discourse. The most vulnerable groups in the world are exposed to this dynamic, which is expected to worsen due to climate change if effective laws and regulations are not created and implemented. Governments must take several necessary steps in response to climate change, including investing in and adapting significant infrastructure to withstand the impact of climate change, managing disaster risks, providing effective land-use planning, and facilitating better information on climate change. Incorporating climate change adaptation into long-term development plans and reflecting these goals in policy and budgeting are necessary to make successful infrastructure investments and prioritise them. Climate change adaptation is primarily about making no-regret development decisions because many of these expenses would have been warranted even in the absence of climate change. Adaptation may be seen as a continuation of good development practices that minimise vulnerability and limit exposure.

Adaptation may have major macroeconomic benefits, even though they are difficult to quantify. In a Vietnamese study, computational general equilibrium modelling was utilised to ascertain that the agriculture sector yielded the most significant adaptation gains, ranging up to 10 per cent of sector value-added relative to baseline in 2050. Furthermore, a notable 27 per cent value was added by the Central Highlands, a region characterised by a high poverty rate and vulnerable population groups. Approximations suggest that Asian cities could face substantial financial costs due to climate change.[3]

This chapter aimed to provide an overview of climate change in the Asia-Pacific region, analyse the geopolitical and geoeconomic impacts of climate change on the region, and examine the adaptation strategies utilised by countries in the area. Climate change is expected to affect numerous industries, and several economies in the Asia-Pacific Economic Cooperation (APEC) region are at risk. In addition to enhancing user satisfaction with climate information, particularly concerning APCC's monitoring data on extreme climate events throughout the Asia-Pacific region, improving cooperation with APEC member economies would also enhance climate forecasting capabilities.

The Climate Change Situation in the Asia-Pacific Region

Climate Change

Long-term changes in temperature and weather patterns are referred to as climate change. These fluctuations may be caused by significant volcanic eruptions or variations in the sun's activity. However, human activity has been the primary cause of climate change since the 1800s, mainly due to the combustion of fossil fuels like coal, oil, and gas.[4] An international response is required since climate change impacts the entire planet. In recent years, combating climate change has been a primary focus of global policy. Climate change is the major environmental element affecting peace and security, and its impacts are predicted to affect the world's most vulnerable communities disproportionately. Climate change has the potential to create new sources of instability, extend unrest to neighbouring countries, and exacerbate low-intensity conflicts currently in place. Despite the UN Secretary-General and General Assembly's recognition of the links between security and climate change, national, regional, and international peace and security strategies continue to overlook the issue riskily.[5]

Asia-Pacific Region

The Asia-Pacific countries are already experiencing the adverse effects of climate change, namely, increased extreme weather events, rising sea levels, and increased resource scarcity. These are likely to act as 'threat-multipliers' to conflict in areas already facing development and security challenges, and where the direct physical effects of climate change will be greatest, leaving many Asia-Pacific countries particularly exposed. The region is home to two-thirds of the world's poor and almost half its intra-state conflicts. It is already facing dire supply and demand challenges, and its populations are more likely to experience natural disasters than any other region in the world.[6]

Furthermore, the Asia-Pacific region has seen a noteworthy increase in the utilisation of renewable energy in recent times, supported by policies that facilitate it, such as targets and subsidies, as well as considerable reductions in the cost of critical technologies like solar and wind. This is due to several advantages: improved energy security, affordability, a smaller environmental footprint, rapid deployment, and simpler energy access. The commercial case for renewable energy is further strengthened by the socio-economic co-benefits it offers. Increased subregional and regional electrical interconnectivity can significantly help large-scale wind, solar, and hydropower exploitation. Since a lot of these resources are cyclical and seasonal, exporting excess electricity across borders to countries facing energy shortages can increase energy security, lower costs, and cut carbon emissions by utilising recently launched renewable energy sources.[7]

Problem

Climate change can affect both environmental and economic security through its impacts on the natural and built environments. The governments that must cope with them face conventional security risks due to those dangers to human security. There is a clear correlation between climate change and environmental security. Variations in precipitation, rising sea levels, and extreme weather can negatively affect freshwater resources and food production in areas that are already vulnerable. Permafrost melting, coastal storms, and riparian flooding affect the built environment. Environmental deterioration and the effects of climate change on food, energy, and infrastructure prices pose threats to economic stability.[8]

Additionally, the United Nations Framework Convention on Climate Change (UNFCCC) has been instrumental in promoting international cooperation to address the problem of climate change.[9] The UNFCCC aims to stabilise Greenhouse Gas (GHG) atmospheric concentrations and provide a long-term solution to the issue of climate change. The principle of common but differentiated responsibility has been the guiding force behind this effort. It is imperative that all countries, both developed and developing, take part in this global solution, even though the actions of significant GHG-emitting economies hold the key to its success.

While reducing Greenhouse Gas (GHG) emissions is crucial in preventing catastrophic levels of global warming, climate change adaptation is equally essential in reducing and limiting the costs of the predictable consequences of GHG emissions that are already embedded in the climate system. Unfortunately, those most vulnerable to the impacts of climate change are often low-income people with limited resources and access to infrastructure, services, and knowledge, making it difficult for them to adapt. Consequently, adaptation is particularly important for developing nations that are striving to combat poverty.

In addition to living in geographically vulnerable places that are prone to natural disasters, they usually work in climate-sensitive industries, including forestry, fisheries, and agriculture. Additionally, their options for changing to different revenue streams are extremely limited. Climate change adaptation should, therefore, be a vital component of any developing nation's development and poverty reduction policies through the creation of adaptive capacity, the implementation of targeted adaptation measures in essential sectors that are vulnerable to climate change, and the provision of aid to the impoverished to cope with its effects.[10]

The Asia-Pacific region is facing an increased risk of adverse effects of climate change due to pre-existing challenges related to poverty, governance, insecurity, and exposure to natural disasters. Despite the region's notable economic growth and progress towards achieving the UN's Millennium Development Goals (MDGs) aimed at eradicating poverty, these challenges pose significant security and environmental concerns that exacerbate the impact of conflicts and natural catastrophes. In this region, where half of the world's population and two-thirds of its impoverished people reside, 1.8 billion individuals survive on less than $2 per day, with a billion of them living on less than $1.25.

In addition, the likelihood of natural disasters striking people in the region is four times higher than that of inhabitants in Africa and several times higher than that of residents in Europe or North America. Additionally, there is a higher likelihood of conflict in the Asia-Pacific. Roughly half of all intra-state battles worldwide in 2008 and one-third of all battle-related deaths in the preceding 25 years occurred in Asia. 'The majority of the recent wars have taken place in Asia: a pattern visible throughout most of the period following the Second World War', according to Patterns of Peace and Conflict, 2009, UCDP and PRIO. Wars in Asia typically endure longer than in other regions, which is another unsettling tendency.

In the Asia-Pacific area, resource-based conflicts are a common occurrence, and the most frequent form of such conflicts is resource conflicts. These conflicts arise when competing interests conflict within or among organisations. The root cause of such conflicts is often the creation and implementation of laws and policies that do not consider the needs of specific user groups. This creates a dangerous dynamic exacerbated by a top-down approach to policymaking, which excludes disadvantaged communities and groups from the decision-making process. As a result, people whose rights and ways of life are routinely disregarded become further weakened. This can trigger a vicious cycle where supply and demand imbalances deepen, leading to poverty and putting communities and states at risk of resource-related violence.

The Impacts of Climate Change in the Asia-Pacific Region

The Asia-Pacific region is known for its resource-based conflicts. However, the data on such conflicts often need to be more accurate and consistent. A study conducted in 2000 revealed several causes of conflict in this region, including natural gas and oil in Afghanistan, timber and heroin in Myanmar, and timber, rubies, and oil in Cambodia. Timor-Leste's oil, Indonesia Aceh's oil, timber, copper, oil, and hydropower in Indonesia-West Papua, copper in Papua New Guinea-Bougainville, and timber and marijuana in the Philippines are other significant sources of conflict. Water systems, being an essential energy source, have the potential to generate hydroelectricity, and their impact on the river systems cannot be ignored. Water scarcity impacts various domains, such as public health, livelihoods, and agricultural productivity, thereby posing a threat to development and security.

The glaciers of the Himalayas, Hindu Kush, and Tibetan Plateau are currently the primary source of water supply for 1.3 billion people residing in key Asian river basins, including the Mekong, Irrawaddy, Indus, Ganges, Brahmaputra, Yangtze, and Yellow River, particularly during the dry season. However, due to the retreat of these glaciers caused by rising temperatures, there appears to be a gradual decline in dry season flows, and water supplies are becoming increasingly unpredictable. With droughts following flooding, water-sharing agreements may also lead to disagreements and, at times, conflicts due to the shifting water flows and levels. Moreover, climate change-induced fish migration may make fishery access agreements irrelevant. Additionally, border disputes over the South China Sea and Exclusive Economic Zones, which extend 200 nautical miles offshore from the coast, may become more intense or take centre stage as a result of the changing coastlines and rising sea levels brought on by climate change.

The agreement between Thailand, China, and Myanmar in 2010 on the potential development of a hydroelectric project in the Salween River has brought to light the dynamics of subregional trade. Although the downstream effects of the project on users have raised concerns with India, China has maintained that the impact will not be significant. In the face of water scarcity, inadequate international water agreements and insufficient evaluation of national water management systems pose a significant threat to national security. Numerous countries in the Asia-Pacific region have attempted to tackle this problem.

Water management is a critical aspect of regional and national governance, both internationally in Association of Southeast Asian Nations (ASEAN) and domestically in Bangladesh. The effectiveness of current approaches has been questioned by ongoing conflicts over water availability, which may ultimately lead to peace or war. According to an analysis by the Institute for International Conflict Research (HIIK) in 2010, resource-related issues were a factor in 12 Asia-Pacific wars that fell under the Asia-Oceania category. Among the nations involved were Timor-Leste, the Solomon Islands, Bangladesh, India, Myanmar, China, Vietnam, Taiwan, Brunei, Indonesia, Malaysia, Pakistan, and the Philippines. Of these disputes, two were overt, four were hidden, and six were of critical severity. However, the study may have yet to consider other resource-based disputes, such as those involving Myanmar, Nepal, or

Fiji. In September 2010, Fiji stated that resolving resource-based conflicts was one of its primary objectives.

Based on a different study conducted between 1980 and 2005, which analysed environmental disputes in the Asia-Pacific region regarding four environmental factors – water, land/soil, fish, and biodiversity – it was observed that many environmental conflicts in the region involved either soil or water. However, it is essential to note that the research did not purport to cover all resource-based disputes in the area. While there are certain commonalities in the data on resource conflicts in the Asia-Pacific region, there are also some significant variations in conflict identification. As a result, the data available for developing comprehensive and focused peace and security-based strategies on existing and potential resource-conflict fault lines needs to be improved.

The challenge of understanding and addressing resource conflicts is further compounded by inadequate reporting, which can be attributed to factors such as corruption, fear, and restricted access to relevant information. This lack of comprehensive information poses a significant hurdle to identifying fault lines in resource conflicts, particularly those that are less visible and may exacerbate environmental stress on a smaller scale. To develop effective mitigation and adaptation policies, access to accurate and timely resource information is essential.

East Asia

Climate change is projected to lead to higher temperatures and increased precipitation in East Asia. This region is particularly vulnerable to natural disasters due to its geographical makeup and the large number of people that could be affected. Deforestation and changes in land usage may exacerbate this situation even further. While some East Asian countries are not solely responsible for climate change, China, the largest emitter of greenhouse gases, significantly mitigates its impact. China's efforts to adapt to climate change will have significant immediate and long-term effects on numerous Asian countries.

The reliance of several East Asian countries on the Mekong River for water poses a potential source of tension as China seeks to meet the water needs of its population. The Mekong River originates in China and provides water to Laos, Myanmar, Cambodia, and Vietnam. Reducing water flow to the end-users and increasing utilisation by countries closer to the source could have detrimental effects on agriculture, water availability, and public health. Therefore, the significance of China's role in addressing the impact of climate change cannot be overstated.

The following case study discusses China's investments in clean energy development, which have led it to become the world's greatest emitter of greenhouse gases. Its neighbouring countries are and will continue to be affected by China's efforts to meet the needs of its rapidly expanding population and economy. As a major player in the global economy, China holds great potential to contribute significantly to regional and global efforts to mitigate and adapt to the effects of climate change. China's growing strength has resulted in new requirements to secure resources to support its population's rising living standards, constituting one-fifth of the world's total population. Due to industrialisation and urbanisation, China's metal consumption share has risen 10-25 per cent since the late 1990s. Furthermore, China is now Saudi Arabia's largest customer, importing half of its oil, and this share is expected to increase to two-thirds by 2015 and four-fifths by 2030. Finally, China is one of the world's largest timber importers, partly due to constraints on domestic forestry activities.

China, which constitutes one-fifth of the global population, is experiencing a rapid decline in its agricultural land area, currently at only 7 per cent The situation is further deteriorating due to increased desertification. Beijing regards the South China Sea as a crucial source of food supply for its citizens, but the country's fish consumption is just over a third of the global average. China's water consumption is expected to escalate in the coming years. However, mounting evidence suggests that climate change could affect the crucial water supplies in the Himalayas and the Mekong River Basin.

In 2006, the Chinese government's first official study on climate change revealed that the country is likely to face severe water shortages in the coming decades. The study also predicted a possible sharp increase in precipitation levels. The agricultural economy is expected to be significantly impacted by the more frequent occurrences of droughts, heat waves, and other extreme weather phenomena. The country's economy has already been adversely impacted by a series of prolonged droughts in recent years despite the disastrous flooding that occurred this summer.

Significantly, China's water challenges are of substantial importance to the entire region. The Himalayan glaciers and Tibetan Plateau, also known as the 'Water Towers of Asia', play a crucial role in feeding seven of the largest rivers in the world, including the Ganges and Yangzi, supplying water to nearly 40 per cent of the world's population. As a result, any disruption in water supplies to China and India from this strategically important region is likely to exacerbate tensions in an already strained relationship. Northern India is home to numerous significant rivers, such as the Brahmaputra, which originates in Tibet, China, and supports millions of people. Last summer, China's decision to build the Yarlung Tsangpo Dam on the Brahmaputra was interpreted by some as indicating that Beijing would divert the river's waters for Chinese farmers' use. This move has raised long-standing concerns in India.

Thailand, Laos, Cambodia, and Vietnam have long complained that China operates on the Mekong without prior consultation. A Bangkok Post editorial accused China of 'Killing the Mekong'; distraught Thai, Laotian, and Cambodian fishermen and farmers have blamed the Chinese dams for reducing fish stocks, cutting irrigation, and disrupting livelihoods.[11] Even while China makes up less than one-sixth of the Mekong's total catchment, it has managed to retain part of the flow during the dry season, which is vital to its neighbours. Following a protracted, severe drought in Yunnan, water levels on the Mekong, whose river basin is the world's largest inland fishing zone, were critically low until the 2010 floods. China is poised to emerge as a significant force in the wind, solar, and hydropower sectors after pledging to more than quadruple its dependence on renewable energy sources by 2020. It is the world's largest manufacturer of wind turbines and is heavily investing in hydropower to fuel its electrical system.

China's more stringent fuel-emission regulations and a stable market for private investments in clean energy technologies put it in a more advantageous position than its main rival, the US. According to some experts, China's long-term goal is to establish leadership and maintain stability while the world moves towards a clean energy future. China is presently the leading producer of parts for renewable energy technologies and plans to invest a considerable sum of USD 1,000 billion over the next decade to meet its ambitions.

China's cooperation with its ASEAN neighbours on environmental issues related to the South China Sea has been hindered by political, cultural, and ethnic differences. Competing claims to strategically important sea areas have led to political confrontations between China and other ASEAN members. However, the increasing frequency of high-level political interactions has led to the emergence of forums that have gradually improved cooperation. China's active political engagement and participation are crucial to this trend. Nevertheless, economic and sovereignty concerns for China remain challenging, and it remains cautious about multilateral forums.

Indonesia's forests contribute to over 75 per cent of the country's greenhouse gas emissions, a situation that has been associated with deforestation, conflict, and corruption. Addressing the links between the forest sector, conflict, deforestation, and climate change is critical. Illicit logging, which has cost the country an estimated USD two billion annually in the past, is a critical factor that connects these two issues. Despite persistent illegal logging, significant progress has been made under President Yudhoyono's leadership. Addressing the negative aspects of the forest sector, particularly corruption, would help minimise the impact of climate change and support peace and security efforts. It is worth noting that Indonesia has the world's fastest pace of deforestation and is one of the top three emitters globally, per a 2007 assessment commissioned by the World Bank and the UK Department for International Development (DfID).

Despite a backdrop of violence and corruption, Indonesia has become a noteworthy example of conflict timber in the forestry sector. The cooperation between Indonesia and the European Union's Forest Law Enforcement, Governance, and Trade (FLEGT) has recognised this connection and established several Conflict Resolution Forums. These forums have full authority to resolve forestry-related conflicts amicably. The current administration inherited policies that may be traceable to the problems with Indonesia's forests.

According to a report published by Associates in Rural Development (ARD) in 2003, conflict timber in six different Indonesian provinces was examined, and it was found that despite variations in the features of conflict timber among provinces, one common theme was observed in all cases. The influence of former colonial governments and/or Suharto's authoritarian rule over Indonesia's forests and resources emerged as the foremost factor. Suharto's dictatorship had strictly regulated resources for its financial gain, with little benefit to the Indonesian people. Threatening each other was used to quell confrontation. Conflicts relating to the wood trade emerged between and within corporations, communities, foreign interests, and different governmental and security force levels after Suharto and with decentralisation. The decentralisation trend, corruption, and the military, political, and bureaucratic weakening have subsequently exacerbated the situation.

South and West Asia

South Asia is a region that is home to half of the world's people with low income. It is highly susceptible to the adverse effects of climate change due to its expanding population, high levels of poverty, exposure to violence, and harsh natural conditions. A catastrophic weather event that affected 50 per cent of South Asia's population between 1990 and 2008 resulted in over 60,000 fatalities and USD 45 billion in damages. A study by the German Advisory Council on Global Change, Climate Change as a Global Security Risk, suggests that climate change in the subregion of India, Pakistan, and Bangladesh could increase the potential for social crises, which are already characterised by cross-border conflicts between India and Pakistan, unstable governments in Bangladesh and Pakistan, and religious fundamentalism.

The Dhaka Declaration on Climate Change and the Action Plan on Climate Change were adopted by the environment ministers of the South Asian Association for Regional Cooperation (SAARC) in 2008. It aimed to address the regional aspects of climate change. It identified critical areas of cooperation, such as adaptation, mitigation, technology transfer, finance and investment, education and awareness, management of impacts and risks, and capacity building for international negotiations. Despite these efforts, concerns about the potential impact of climate change on regional and national security persist. The conversation over climate change generally disregards peace and security issues.

The issue of water supply has become a pressing political concern in South Asia, where rivers serve as a critical resource for both upstream and downstream countries. The melting of the Himalayan glaciers, the primary water source for many of the region's vital rivers, is a direct consequence of climate change. The water supply interruption to these rivers could have far-reaching impacts on the subregion and its inhabitants due to their reliance on these water sources. This issue could be especially challenging when combined with predictions of more frequent and severe weather events, rising temperatures, altered precipitation patterns, growing populations, and increasing demand for water.

The distribution of freshwater is unequal, and this is exacerbated by pollution and changes associated with climate change. This has been both a source of past cooperation and tension. For example, the building of the Baglihar Dam over the Chenab River in Indian-administered Kashmir has increased tensions between India and Pakistan. Rising sea levels will also have an impact on those who live along Bangladesh's, India's, and Pakistan's coasts, perhaps leading to mass migration. The Indus River Basin, which provides millions of people in Pakistan and India with their main water source, is essential. Dams and canals have been constructed to generate electricity; however, this has also caused some areas of the Indus River Basin to dry out and encouraged cross-border migration. In addition to being significant economically and religiously, the Ganges River Basin is a cause of contention between Bangladesh and India.

Pacific Scientific Research has advised several small island governments to prepare for potential humanitarian disasters resulting from increased extreme weather events and rising sea levels. While a significant portion of migration is domestic, a considerable amount of migration related to dissolving nations may become transnational due to long-term uncertainty and increasing greenhouse gas emissions. In its submission to the UN Secretary-General's report, Climate Change and its Possible Security Implications, the PSIDS noted that they were dealing with multiple challenges, such as rising sea levels, changing weather patterns, natural disasters, soil erosion, and loss of coral reefs. These factors have adversely affected public health, physical and social infrastructure, loss of life and livelihoods, migration, loss of islands, territorial integrity, sovereignty, legal rights, conflict and unrest, and sociocultural impacts.

In addition to displacement and migration, several Pacific Island nations have experienced heavy inundation of densely populated coastal areas, loss of freshwater, failure of agriculture, and other consequences of saltwater intrusion. As mentioned earlier, these have led to severe food and water shortages, forcing people and towns to flee to Vanuatu, Micronesia, Papua New Guinea, Tuvalu, and the Solomon Islands. These shortages have been exacerbated by the effects of climate change, such as recent sea-level-rise-related pollutants and salt deposits, which have had a detrimental impact on food and water resources. The water scarcity in Kiribati has reached a point where the country is on the verge of declaring a state of emergency.

Geopolitical and Geoeconomics

The Belt and Road Initiative (BRI) of China, a regional cooperation programme designed to promote economic growth, investment, and commerce among 68 Eurasian nations, has the potential to impact regional climate initiatives significantly. While the primary objective of the BRI is to foster regional economic cooperation, its emphasis on energy and infrastructure could affect the mitigation and adaptation efforts of the participating nations. The economies of BRI countries are typically high in carbon, with per capita carbon dioxide (CO_2) emissions averaging 6.13 tons annually, significantly higher than the global average of 4.17 tons per capita. Furthermore, the economies of BRI countries emit 0.7 kg of CO2 emissions per USD of GDP, compared to the global average of 0.44 kg of CO2 per USD of GDP, indicating their higher carbon intensity. Without appropriate laws, accelerating infrastructure development in the BRI countries could lead to even higher carbon emissions.

China may consider relocating its surplus manufacturing capacity to countries participating in the Belt and Road Initiative (BRI). The country's carbon-intensive sectors, including rolled steel, electrolytic aluminium, cement, plant glass, crude steel, and thermal power, currently have a surplus production capacity, which is being further exacerbated by the expansion of its capacities in thermal power, electrolytic aluminium, and rolled steel. There are concerns that the BRI may inadvertently increase carbon emissions, as one of its objectives is to increase global demand for Chinese goods from these carbon-intensive industries. China has been emphasising the significance of the 'green BRI' due to its realisation of potential environmental and climate risks.

As the world's biggest greenhouse gas emitter, China has committed, under its Nationally Determined Contribution (NDC) under the Paris Agreement, to guarantee that its carbon emissions will peak in 2030. Reducing leakage and assisting China in leading on climate change worldwide will be achieved by including carbon emission regulations in the BRI. On the other hand, China wants to optimise its industrial structure to address the issue of surplus capacity. Examples include closing outdated manufacturing sites and providing subsidies for green sectors. By decarbonising China's outbound trade and investment, the Belt and Road Initiative (BRI) has the potential to become a significant platform for supporting low-carbon development and raising ambition for mitigating climate change, since China is steadily developing a competitive edge in green technology and goods.[12]

Policy APEC

In 2007, the UN Security Council held a debate on 'The possible security implications of 47 climate change', attended by over 50 member states, marking it as the only UN body specifically mandated to address issues that may threaten international peace and security. During this debate, climate change and security were brought to the forefront, revealing a distinct division between governments that favoured and those that opposed the Security Council's attention to the issue. The sharp differences and concerns raised during the discussion have limited further attention from the Security Council. However, this may change as Germany, a 2-year member of the Security Council since January 2011, has stated that it views climate change as a security concern that requires immediate attention. The country's ambassador to the UN made it clear that they aim to bring the matter before the Security Council, despite the possibility of objections from some members who argue that it has no bearing on the council's authority.

However, the Security Council has actively participated in debates on thematic threats to peace and security. The nature of modern security threats is significantly different from the conventional interstate military threats prevalent during the Cold War era. In this regard, the United Kingdom has circulated a concept paper titled 'Energy, Climate, and Security' before the 2007 Security Council climate change discussion. The discussion will primarily focus on 'the security implications of a changing climate, including its impact on potential drivers of conflict'. The paper explicitly outlines six critical effects of climate change that could threaten global peace and security: border conflicts, migration, resource scarcity, social unrest, and humanitarian disasters.

The concept note presented a view that 'cumulative impacts of climate change could exacerbate these drivers of conflict, and particularly increase the risk to those states already susceptible to conflict', instead of attributing direct causality of conflict to climate change. During the discussion, Russia, South Africa, Brazil, China, and India expressed their opposition to the concept, citing that it interfered with the roles of other UN agencies and that climate change was a sustainable development matter, not a security issue. Moreover, there was concern that any action taken by the Security Council to address climate change would place uniform and broad obligations on all nations, disregarding the origins and consequences of the phenomenon and violating state sovereignty.

Concerns have been raised regarding the negative socio-economic impact that may arise from the expenses required for climate change mitigation, which could potentially halt growth. According to a study on the legal dynamics of climate change and the Security Council, the political will of individual members will be a determining factor in the ability of this organ to address the challenges posed by climate change. The Security Council has been vested with the legal authority to address the causes and consequences of this threat. Ultimately, the most crucial factor is not whether the members choose to exercise this power but rather whether they do so in an effective and impactful manner.

International cooperation is imperative for mitigating the effects of climate change, which is a global threat to resource and security concerns. However, any international entity still needs to comprehensively address the issue of resource-based conflicts and climate change-related conflicts. Nonetheless, there are several forums for addressing the remaining 54 climate change-related challenges, including the UN frameworks, which may be employed to address conflict prevention and resolution. With nearly universal membership, the United Nations Framework Convention on Climate Change (UNFCCC) is currently the most widely ratified climate change treaty. However, the UNFCCC's recommended measures are purely voluntary and lack legal force, with several countries opting not to follow them. The UNFCCC acknowledges that the global nature of climate change necessitates the participation of all countries and the broadest collaboration in an appropriate and effective international response. Unfortunately, weak collaboration and international response have been insufficient to address the scope of the risks posed by climate change, and the UNFCCC needs to be mandated to address the consequences of climate change that threaten international peace and security.

The Kyoto Protocol, which is a part of the Framework Convention, came into effect in 2005 after initially being made available for signature in 1998. The protocol's greenhouse gas objectives are legally binding on all parties, although they are only sometimes accepted or constrained over the long term. During the 2009 climate talks in Copenhagen, rich and developing nations discussed the most effective approach to tackling climate change, with no shortage of facts. During the conference, the countries known as BASIC (Brazil, South Africa, India, and China) collaborated to demonstrate their preference for forming alliances and engaging in negotiations based on economic alignment rather than proximity.

At the climate talks in Cancun, significant breakthroughs still needed to be achieved. However, the conference did surpass its expectations by agreeing to several new and modest agreements. These agreements include reducing deforestation and forest degradation and creating a new billion-dollar fund for 'long-term' climate finance from developed to developing nations. The fund will be managed by an impartial board, with the World Bank serving as the trustee. Additionally, the accords included the establishment of 230 new boards, organisations, initiatives, and financing agreements, all of which may pave the way for the UN process to move beyond the flawed Kyoto Protocol and towards new targets. To foster international agreement on climate change, the G-20, which includes the two most developed and fastest-growing economies in Asia, along with ROK, has begun to play a role in 'trying to move toward agreed goals and their implementation and seeking to establish broad aggregate measurements and equally broad commitments on both emissions and financing', among other international initiatives.

The Adaptation Strategies of Countries in the Asia-Pacific Region to Climate Change

Policy APEC

The Asia-Pacific region is at the forefront of the impacts of climate change and is uniquely positioned in global efforts to manage climate change. The region is suffering greatly from rising temperatures, rising sea levels, and extreme weather events brought on by climate change. These factors damage the region's natural resources, economy, and physical assets and worsen developmental issues like poverty, food and energy security, and health. Climate change may push over 100 million people in the area into extreme poverty by 2030 if climate change is not addressed via development that is focused on mitigating its effects. This would reverse the decades long progress made in reducing poverty. Simultaneously, the area is responsible for 53 per cent of global emissions, and as many of the region's economies are growing rapidly, this contribution will continue to rise in the absence of significant policy changes.[13]

Recent estimates show that climate change is expected to significantly impact the Asia-Pacific region's growth. Failure to implement climate action could decrease the region's GDP by up to 3.3 per cent by 2050 and 10 per cent by 2100 compared to the base case. Moreover, naturally occurring calamities impose a more significant financial burden on the region. Between 1970 and 2015, the annual damage to property, crops, and animals due to catastrophes increased from USD 52 billion to over USD 523 billion. Bringing the region down to a two °C scenario, as opposed to business as usual, is estimated to cost 0.1 per cent of GDP annually or 4 per cent by 2050. The co-benefits of taking climate action offset many of the costs associated with reducing emissions, while advanced technologies offer the possibility of lower abatement costs.

The implementation of robust climate change policies can be well-supported by fostering collaboration across regions. Regional cooperation is an essential strategy for effectively addressing various issues, such as harmonising carbon pricing initiatives, exploring potential market linking, developing Internationally Transferred Mitigation Outcomes (ITMOs), promoting technology cooperation, implementing Sustainable Development Goals (SDGs), and raising climate ambitions. Notably, leveraging sub-national networks, including businesses, civil society, and cities, can provide a platform for achieving these objectives.

Adaptation Strategies

Rising sea levels and frequent catastrophic weather events are two direct consequences of climate change that vulnerable communities dependent on the most impacted sectors will face in the Asia-Pacific region. To avoid resource-based conflicts in the context of climate change, state-based efforts must take this fact into account. Therefore, national governments should prioritise the following:

(1) It is crucial to consider IPCC data and forecasts while incorporating environmental and security aspects into development, conflict prevention, and climate change plans, especially in NAPAs. This will ensure that cross-sector approaches are taken toward the

MDG targets. To ensure the participation of vulnerable people, this should involve addressing several intra-state resource custody chains and cross-sector interdependence and mainstreaming security issues into Disaster Risk Reduction strategies (DRRs).

(2) Identifying and resolving existing resource-conflict issues, both high- and low-intensity, and anticipating future conflicts is paramount. Developing comprehensive strategies for preventing conflicts and addressing issues with internal governance, land tenure, and property rights is also important. One way to solve resource-based conflict prevention and resolution oversight difficulties and shortcomings is to design conflict prevention strategies in collaboration with NGOs, CSOs, and vulnerable groups.

(3) Ensuring that there are no 'one-size-fits-all' approaches to resolving resource conflicts or adapting to climate change is crucial. This can be achieved by ensuring that strategies for adapting to climate change are participatory, context-specific, and pro-poor, prioritising community empowerment and ownership and strong representation of vulnerable groups.

(4) Immediate action must be taken to reduce vulnerabilities, support good governance, and advance sustainable human development by managing energy, water, land, and forest resources. Additionally, identifying potential economic and peacekeeping benefits, such as increased employment, from strategies for mitigating and adapting to climate change can be valuable.

In order to ensure the provision of financial and other assistance for resource-conflict mitigation techniques, the international community must consolidate and prioritise systems for preventing resource conflicts. In this regard, it is imperative to expand on UN General Assembly Resolution 63/280, which called on UN agencies to give more attention to this issue while recognising the connections between climate change and security. The following steps must be taken to resolve this issue:

(1) The UN Security Council must acknowledge that climate change may threaten peace and security. Consequently, the council should revise its mission to include environmental security as a critical subject of conflict prevention. In addition, it should consider creating a subsidiary organisation that would play a crucial role in coordinating security issues related to climate change and collaborate closely with key stakeholders, such as ASEAN, SAARC, PIFS, the IPCC, ECOSOC, OCHA, ISDR, UNHCR, IOM, WTO, and the World Bank. The UN Security Council should address broad problems related to security and climate change and urgent and regional concerns while upholding the principle of common but diverse responsibilities.

(2) The UN Department of Political Affairs (UNDPA) and the UN Peace Building Commission (UNPC) must prioritise and finance vital responsibilities in conflict prevention. They must ensure that initiatives for peacebuilding and conflict prevention consider resource-based conflicts and climate change.
(3) The UN Environment Programme (UNEP) must be given an independent budget, sufficient funding, an enlarged mandate, and support for the launch of initiatives that guarantee resource- and environmentally conscious peacebuilding and conflict prevention and resolution initiatives in the context of climate change. The UNEP's Division of Early Warning and Assessment (DEWA) and Post-Conflict and Disaster Management Branch (PCDMP) can handle resource-based conflicts. This aligns with the UN-specialised agency's upgraded status, recognised by international environment ministers in February 2011, following the need for more attention to environmental challenges, particularly considering climate change.
(4) The UNEP and UN Development Programme (UNDP) must enhance current coordination and collaboration between them to address the symbiosis between development, security, and the environment, especially at local and regional levels, as demonstrated by the Poverty-Environment Initiative (PEI).
(5) The UNEP, the UNDP, the Food and Agriculture Organization (FAO), UNDPA, the United Nations High Commissioner for Refugees (UNHCR), and the International Organization for Migration (IOM) must enhance agency-to-agency cooperation in resolving resource-based disputes and potential implementation considering climate change.
(6) The World Bank must improve its Conflict Prevention and Reconstruction Unit (CPRU) by addressing resource-based conflicts, how to prevent them, and how climate change effects may cause a rise in events. This could be included in and supported by World Bank Strategic Environmental Assessments (SEAs) data.
(7) The UNISDR must ensure that low- and high-intensity resource-based conflicts are addressed in the national and regional platforms that are currently in place, such as the Asian Partnership of Disaster Reduction (IAP). This is because forced migration, scarcity dynamics, and opportunism frequently accelerate the links between natural disasters and insecurity.

The Paris Agreement has successfully transformed global climate objectives into a widely accepted agenda with well-defined objectives.[14] Nevertheless, many nations still need help in financing the estimated costs of reducing emissions while also grappling with the difficulties and hazards associated with implementing long-term policies to support the transition to low-carbon growth. This is evidenced by the difference in emissions between current and necessary climate commitments to achieve the Paris goal. By 2030, it is expected that Asia-Pacific countries will need to increase their current levels of reduction ambition four-fold to bridge this gap.

Among all regional organisations, APEC possesses the institutional framework, experience, and political will to unite its members and galvanise them to tackle the threats to economic and environmental security caused by climate change and achieve increased regional cooperation. These approaches showcase the varied paths taken by APEC member economies to adapt to the consequences of climate change. By addressing vulnerabilities across multiple sectors and promoting collaboration, APEC aims to enhance resilience and ensure sustainable development in the face of climate change.

Conclusion

The Asia-Pacific region faces a daunting challenge in addressing climate change. Given the region's diversity and the varying growth stages of its constituent countries, there is no one-size-fits-all solution to the numerous issues at hand. Many states, particularly the small island nations, emerging countries, and the least developed, are grappling with significant challenges related to poverty, infrastructure, and inclusive growth. Climate change is expected to exacerbate these issues and create new ones. As sea levels rise, resources deplete, and extreme weather events become increasingly frequent, the already precarious conditions in the region may deteriorate further. In a discussion on energy, security, and climate change at the UN Security Council, former UN Secretary-General Kofi Annan warned that environmental degradation could destabilise conflict-prone regions, particularly when access to scarce resources is inequitable or politicised. Given these risks, the Asia-Pacific region must make difficult choices to mitigate the impact of climate change.

The region's challenges cannot be addressed with a 'one-size-fits-all' approach due to the varying levels of development among countries in the area. Governments in the region, especially those in tiny island states, least developed nations, and landlocked developing nations, face significant obstacles in combating poverty, addressing infrastructure needs, and achieving equitable development. Climate change will exacerbate these challenges, which will also give rise to new issues. Despite the considerable economic progress made in the region, which has lifted millions of people out of poverty and transformed the area, it has a high carbon footprint and needs to account for the future costs of climate change.

As a result, the Asia-Pacific region could potentially introduce new laws, incentives, and policies to transform its economy in favour of low-carbon growth. Practical regional cooperation is a vital tool that connects global, national, and local climate action. The Economic and Social Commission for Asia and the Pacific (ESCAP) is well-positioned to assist its member states in developing these policy frameworks through a range of initiatives for regional cooperation. These initiatives include sub-national networks, improved data collection, enhanced capabilities for science, technology, and innovation, collaboration on shared vulnerabilities, harmonisation and integration of multi-hazard early warning systems, integration of carbon markets, green finance, and private sector initiatives, and sharing of regional knowledge and best practices. Through its multi-sectoral strategy, norm-setting, and international platform, ESCAP can help its members establish resilient, low-carbon economies, ensuring their continued prosperity in a world with carbon restrictions.

Notes

1 APEC, 'https://www.apec.org/About-Us/About-APEC/Fact-Sheets/Climate-Change', December 2021. (accessed January 22, 2023) The Asia-Pacific Economic Cooperation (APEC) is a regional economic forum that promotes cooperation and sustainable development among its member economies around the Pacific Rim. APEC was established in 1989 with the primary goal of advancing economic growth, trade, and investment in the Asia-Pacific region. The organisation operates on the principles of open dialogue, economic cooperation, and non-binding agreements.
2 World Bank, 'Managing the Impacts of Climate Change on Poverty', Washington, D.C.: World Bank, 2015.
3 World Bank, 'Climate Risks and Adaptation in Asian Coastal Megacities: a synthesis report', Washington, D.C.: World Bank, 2010b.
4 United Nations, 'https://www.un.org/en/climatechange/what-is-climate-change', (accessed January 22, 2023)
5 Alice Blondel, Climate Change Fuelling Resource-Based Conflicts in the Asia-Pacific, Asia-Pacific Human Development Report Background Papers Series 2012/12, p. 8.
6 Alice Blondel, Climate Change Fuelling Resource-Based Conflicts in the Asia-Pacific, Asia-Pacific Human Development Report Background Papers Series 2012/12, p. 8.
7 ESCAP, The Economics of Climate Change in The ASIA-PACIFIC REGION, 2016, p. 13.
8 J. Scott Hauger, 'Climate Change and Environmental Security in the Asia-Pacific Region: A Role for APEC?', Affairs Journal 9.2 (2014): 131–135.
9 United Nations, 'United Nations Framework Convention on Climate Change', 1992 conveng.pdf (unfccc.int) https://unfccc.int/resource/docs/convkp/conveng.pdf, (accessed January 22, 2023)
10 Asian Development Bank, 'The economics of climate change in Southeast Asia: a regional review', Mandaluyong City, Philippines: Asian Development Bank, 2009, pp.3–4.
11 Alice Blondel, 'Climate Change Fuelling Resource-Based Conflicts in the Asia-Pacific', Asia-Pacific Human Development Report Background Papers Series 2012/12, p. 97.
12 Ziting Huang and Junjie Zhang, 'Developing Non-Market Approaches through Regional Cooperation Platforms', in International Cooperation in East Asia to Address Climate Change, edited by Robert N. Stavins and Robert C. Stowe, February 2018, pp. 66–67.
13 ESCAP, 'The Economics of Climate Change in the ASIA-PACIFIC REGION', 2016, p. IV.

14 United Nations Climate Change, 'The Paris Agreement: What is the Paris Agreement?', The Paris Agreement | UNFCCC, https://unfccc.int/process-and-meetings/the-paris-agreement (accessed January 22, 2023).

References

APEC. (2021, December). *Climate change.* https://www.apec.org/About-Us/About-APEC/Fact-Sheets/Climate-Change

Asian Development Bank. (2009). *The economics of climate change in Southeast Asia: A regional review.*

Blondel, A. (2012). *Climate change fuelling resource-based conflicts in the Asia-Pacific* (Report Background Papers Series 2012/12, p. 8). Asia-Pacific Human Development.

ESCAP. (2016). *The economics of climate change in the Asia-Pacific Region.* https://www.unescap.org/kp/2016/economics-climate-change-asia-pacific-region

Hauger, J. S. (2014). Climate change and environmental security in the Asia-Pacific region: A role for APEC? *Affairs Journal, 9*(2), 131–135.

Huang, Z., & Zhang, J. (2018). Developing non-market approaches through regional cooperation platforms. In R. N. Stavins & R. C. Stowe (Eds.), *International cooperation in East Asia to address climate change* Cambridge, MA: Harvard Project on Climate Agreements, (pp. 66–67).

United Nations. (1992). *United Nations Framework Convention on Climate Change.* https://unfccc.int/resource/docs/convkp/conveng.pdf

United Nations. (n.d.). *What is climate change?* https://www.un.org/en/climatechange/what-is-climate-change

United Nations Climate Change. (n.d.). *The Paris agreement. What is the Paris agreement?* https://unfccc.int/process-and-meetings/the-paris-agreement

World Bank. (2010). *Climate risks and adaptation in Asian coastal megacities: A synthesis report.*

World Bank. (2015). *Managing the impacts of climate change on poverty.*

4 Climate Changes and Adaptation Measures in the South Pacific

Edmund Rice Komboli

Introduction

The purpose of this chapter is to inform people around the world about the real impact of climate change on small Pacific Island countries and the measures taken to address this issue of climate change going forward. It is important to protect the shoreline, whether it be through replanting new trees, building seawalls, or doing land reclamation. Small Pacific Island countries such as Tuvalu are a good example, as they are currently working on land reclamation projects to take back what was already lost to the effect of climate change. Coastlines are important as they are the first defenders against the impact of climate change. People in the Pacific have been informed and are aware not to cut trees along the coastlines and are also encouraged to plant more trees like mangroves and other trees that can help withstand the effects of climate change.

Mangroves play an important part in the ecosystem, not only supporting the marine resources that people of the Pacific region depend on for food and income, but also provide shoreline protection. Coastlines are a place for recreation, shared cultural heritage, and many other benefits, all of which are at risk from climate change.

The shallow ecosystems and productivity of mangroves, seagrass beds, coral reefs, and inter-reef seabeds are critical natural assets for food production, food security, cultural and recreational activities, and livelihoods for many people in the Pacific Islands countries. They also provide important ecosystem services in protecting the coast against storm surges and in the production of carbonate sands and debris to nourish beaches and maintain islands. These ecosystems are easily damaged through reclamation, drainage, pollution, and destruction of critical habitats for fish and other food species. 'Once destroyed, these ecosystems are not readily or cheaply restored or replaced (Kenchington, 2009)'.

> The vital commercial and national assets, essential infrastructure, and population of most Pacific Island countries lie in the coastal zone. Given that most of the region's population is settled in coastal areas, changes in population density combined with new technology

DOI: 10.4324/9781003586869-5

> and changing development priorities have had a significant impact on coastal environments in the last decade.(SOPAC, 1994)[1]

The coastal processes are hydraulic and sedimentary processes driven by tides, currents, waves, coastal winds, and tsunamis. Forces exerted by wind and water on the ocean floor and shore face drive currents, move sediments, erode exposed bedrock, and shape the coastline, estuaries, and the nearshore seabed.

Most Pacific Islands' coastal zones are now abandoned, and ecosystems shift landward. This choice can be motivated by excessive economic or environmental impacts of protection. In the extreme case, an entire area may be abandoned. Accommodation implies that people continue to use the land at risk but do not attempt to prevent the land from being flooded. The options include erecting emergency flood shelters, elevating buildings on piles, converting agriculture to fish farming, or growing flood- or salt-tolerant crops. Protection involves hard structures such as seawalls and dikes, as well as soft solutions such as dunes and vegetation, to protect the land from the sea so that existing land use can continue.

Conceptual Background of the Theme

Papua New Guinea has an ocean area of over three million square kilometres and a total coastline of 17,110 kilometres. The coastal resources are used by many diverse interests, from subsistence to commercial harvesting. It is necessary to protect, conserve, and manage the natural resources within the coastal zone to ensure that the benefits are sustained and available for the future. The shoreline beaches, mangrove forests, seagrass beds, coral reefs, and the open ocean are all connected. This marine ecosystem provides a supportive habitat for many valuable marine resources. An understanding of marine ecosystems is increasingly important as the sustainable management of valuable marine resources in PNG is challenged by the demands of development, population growth, and adaptation. These consequences pose urgent challenges, including significant impacts on the economy, environment, and livelihood of coastal Papua New Guineans. There is an imperative need to develop people's knowledge and understanding of the marine environment, climate change impacts, and possible adaptation measures appropriate for PNG's coastal areas.

A study by Mcleod et al. (2019) highlights that Pacific Islanders are taking a prominent role in addressing climate change at the local level by executing strategies aimed at enhancing the resilience of communities and ecosystems. This region serves as a crucial arena for the refinement of adaptation measures on a significant scale. The Pacific Islands boast unique biodiversity, with species exclusive to this area, and exhibit remarkable diversity in terms of ecosystems, geography, and demographics. Inhabitants of the Pacific Islands have historically coped with natural environmental impacts over millennia, adapting their practices to accommodate fluctuations in the environment.

Despite the acceleration of environmental and climatic changes, numerous communities are adopting climate-smart agriculture and reviving traditional methods that incorporate drought-resistant species. Additionally, they are harnessing the benefits of nature, such as using seaweed as compost to enhance soil fertility, employing palm fronds to shield plants during droughts, and implementing vegetation planting to mitigate flooding and erosion along coastlines.

Importantly, in the specific context of the Pacific Islands, the coral reef ecosystems, essential for coastal defence, sustenance, and tourism, are currently experiencing substantial adverse effects due to climate change. The rise in sea temperatures is anticipated to intensify tropical cyclones and other precipitation events, worsening the already significant economic and societal consequences associated with such weather phenomena (Wier & Orcherton, 2017).

A study by Lal (2004) reports that in a warmer atmospheric context, occurrences of intense rainfall are anticipated, potentially leading to severe flash floods. Within mountainous landscapes, these extreme precipitation events bear geomorphological significance, as they have the potential to induce widespread slope failures and landslides. The impacts of climate change may significantly alter the dynamics of hydrological systems, erosion processes, and sedimentation in certain regions of Small Island Nations.

Climate change, particularly the escalation in temperature and inconsistent precipitation patterns, has played a significant role in the proliferation of water-borne and vector-borne diseases. The Pacific Islands are also grappling with numerous noncommunicable diseases. According to data from the World Health Organization, the region has experienced over 40 significant outbreaks of infectious diseases since 2012. Of notable concern is the association of a majority of these outbreaks with climate-sensitive diseases such as dengue, chikungunya, and Zika viruses. As an illustration, the anticipated consequences of climate change include an elevated susceptibility to infectious and vector-borne diseases, encompassing diarrhoea and cholera in Kiribati and malaria, dengue, and lymphatic filariasis in Vanuatu.

Acknowledging the crucial role of mangroves in both climate change mitigation and adaptation, along with the potential of blue carbon initiatives to facilitate mangrove management, the Fijian Government has recognised the imperative to counteract the prevailing decline in mangrove cover. The aim is to preserve and sustainably oversee mangrove ecosystems, incorporating the valuation of their ecosystem services into national climate strategies and mechanisms (MoE, 2018). For instance, Fiji has devised a Mangrove Management Plan within the framework of the 2013 Mangrove Ecosystems for Climate Change Adaptation and Livelihoods Project; however, the formal implementation of this plan is pending. Education is the most effective means of achieving such awareness. It is the intent of PNG's 'Vision 2050' to transform its people and nation by reforming mindsets and attitudes for a strong and better future (Cameron et al., 2021).

There are many threats to the shoreline because of direct exposure to the activities of people. The land and sea are linked by coasts, rivers, winds, and humans. Therefore, it is not surprising that many of the activities that occur on land impact marine ecosystems. Deforestation, pollution, development, and overfishing caused by human activities harm marine ecosystems, including the shoreline. Natural events such as storms and seasonal high tides are usually short-term; they occur and end reasonably quickly. Marine environment ecosystems, including shorelines, have shown a remarkable capacity to gradually recover from the impacts of natural threats and events. Sometimes humans take action to reduce the risk of or in response to damage by natural threats, e.g., building protective seawalls or growing sand-stabilising plants such as mangroves. Human activity impacts and threats to the shoreline often occur and continue without realisation. This made it more difficult for an ecosystem such as the shoreline to recover as it does not get an opportunity to re-establish a natural balance. Examples of threats to the shoreline ecosystem as a result of human activities include both animal (e.g., piggeries) and human sewage waste. If not properly disposed of, it will pollute marine ecosystems such as the shoreline. Humans then harvest and eat marine animals contaminated by raw sewage. Shorelines also suffer from rubbish and litter, either from people discarding it along the coast, from inland rivers carrying it to the shoreline, or from items washed ashore from the ocean. This litter harms the natural habitat of the main animals that are part of the shoreline ecosystem. Plastic litter is very harmful as it pollutes the ecosystem and kills animals. Major pollution such as oil spills destroys entire shoreline habitats and their resident animals.

In fact, coastal population increase has placed a strain on shoreline ecosystems due to over-harvesting of edible animals including shellfish and fish near shorelines. Larger animals are often the main reproductive sources of a species' population, so when they are removed, reproduction for the species slows and this negatively impacts the whole population. This affects other animals such as shorebirds as the shoreline food-web balance is disturbed. The demands of an increasing population and a developing economy too often mean that industrial, agricultural, and urban development projects do not have effective environmental protection mitigation plans in place. As a result, ecosystems such as shorelines suffer irreversible damage which adversely affects the animals that rely on these ecosystems. Shoreline projects often involve dredging of sand and construction that devastates the natural habitat. The negative impacts on the shoreline are associated with deforestation and major agricultural and livestock commercial projects, including waste, fertiliser, pesticides, and sediment run-off. Planet Earth, the land, the ocean, plants, animals, and humans are all significantly negatively impacted by the severe weather patterns associated with climate change. Worse, it is clear that climate change is poisoning the open ocean through acidification, so it is inevitable that the shoreline ecosystems, which are the link between the ocean and the land, will also be negatively impacted. Climate change is threatening the continued existence of life on Earth. Scientists have proven that man-made climate change is gradually producing conditions on Earth very similar to factors that preceded and caused natural catastrophic extinction events of plant and animal life both in the oceans and on land in the past.

Adapting to Climate Change and Threats

Rising sea levels and more frequent severe storms have forced some island communities to move elsewhere. Adaptation responses for PNG coastal communities include relocating to higher ground, building protective breakwaters and seawalls, raising houses on poles, and planting mangroves. Where seagrasses have been degraded, destroyed, or threatened, it is possible to respond. There have been successful projects to protect seagrass beds by making areas protected 'tambu' zones from harvesting. Seagrass restoration projects involving replanting of clumps of seagrass have succeeded, although the growth rate is slow and success is conditional on removing the factors that cause degradation in the first place, such as pollution, excess sedimentation, or harmful human activity. Stopping people from walking on and harvesting in seagrass beds will help protect the habitat. PNG coastal communities can set up and monitor Marine Protected Areas (MPAs) to protect seagrass beds. Protecting a seagrass bed allows many marine animals that use the habitat to lay eggs without disturbance from human activity. The hatched larvae eat algae on the seagrass leaves and grow into juveniles, which later, as they get bigger, move to nearby mangroves and coral reefs. A protected seagrass bed then is the source of new stock which grows, moves to other areas, and may then be harvested.

The government is encouraging people living along the coastline to plant more mangroves and other trees that can help withstand the effects of climate change or move to higher ground. The government has also encouraged villages on the low-lying atolls to move to the mainland. PNG has already experienced climate change refugees in the Carteret Islands, one of the islands of the Autonomous Region of Bougainville. The Climate Change and Development Authority (CCDA) recently introduced a new concept to local villages along the coast of Port Moresby to start planting coral. It is planning to implement this in other maritime provinces in Papua New Guinea together with new initiatives of establishing Marine Protected Areas (MPAs). However, more emphasis is on growing mangroves.

Importance of Mangrove Ecosystems

Mangroves are flowering shrubs and trees that have adapted to living close to sandy and muddy shorelines in conditions where other land plants cannot survive. Before human activity depleted them, mangroves existed in the intertidal zone along 75 per cent of tropical shorelines, which is an indication of just how important they are as a marine ecosystem. The importance of mangrove ecosystems is measured in terms of the ecological, community, and economic values that these forests provide. Many marine animals start life in the shelter of the mangrove habitat until they are big enough to move to other ecosystems such as coral reefs. The ecological importance of mangroves is reflected in the fact that many are now protected areas or parks, and rehabilitation replanting projects are being conducted where mangroves have been damaged.

Mangroves are made up of different types of shrubs and trees, which are land plants but have developed salt tolerance and grow where saltwater mixes with freshwater. There are 44 different mangrove species in PNG.

Different species of mangroves have developed different adaptations to grow in various zones between the land and the sea. Some prefer high salinity close to the sea (front zone), while others prefer less salinity (middle zone), and other species dominate an area; they are called a 'stand', not a forest.

Although different mangroves have different levels of salt tolerance, they all need some fresh water to grow, even those that live in the intertidal front zone and get saltwater twice each day. Most mangroves get rid of salt through salt-excretion glands in their leaves, or the salt builds up in a leaf until the leaf simply drops off the tree to become part of the decaying material on the seafloor.

Mangroves prefer sheltered areas with soft sandy or muddy sea floors. The specialised spreading roots trap mud and sediments that wash off the land during heavy rainfall. By acting as a protective buffer zone, mangroves stop the sediment mud from ending up on top of coral reefs and seagrass beds and killing them. Mangroves are true 'partner ecosystems' with seagrasses and coral reefs.

Mangroves are breeding, nursery, and feeding grounds for many fish and other marine species, as well as land animals. Most tropical fish begin life in or near mangroves, where they are protected from predators (bigger fish) and have plenty of food available. Many burrowing marine animals, such as crabs, shrimp, snails, and worms, live in mangroves. Encrusting animals such as oysters live attached to mangrove roots. Shoreline birds come to the mangroves to hunt and feed on the many small animals living there. Land reptiles, including lizards, frogs, and snakes, hunt for food in mangroves.

Mangroves are an important natural protective barrier for coastal areas against erosion and damage from wind and storm surge waves. The spreading root systems hold and stabilise the soft seafloor. As the mangroves enable sediment to build up, along with dead leaves, other land plants can grow. Mangroves drop fully developed seeds onto the substrate most suitable for them to grow in.

Mangroves are evergreen ecosystems that protect coastlines and promote marine life for many species. Like seagrasses, they also support a high diversity of other land plants with similar adaptations for living in the harsh conditions of the shoreline environment, especially in the intertidal zone.

Mangroves have become valuable sites of interest for eco-tourism. In some places, mangroves were being bulldozed for development purposes but are now protected tourist attractions with boardwalks built through them. They are increasingly recognised as valuable resources for environment and ecological education, community cultural traditions including medicines, and recreational activities for nature lovers such as bird watchers.

Given the small size of their economy and their government priorities every year, these countries are unable to put their resources into addressing the issues of climate change. They rely mainly upon countries like Australia and New Zealand to counter-fund their annual budgets so additional funds can be allocated to addressing the issue of climate change in their respective countries. Most of the direct funding comes from international organisations and NGOs who support them, especially when countries are affected by natural disasters like cyclones or tsunamis. Many small Pacific Island countries that cancel their diplomatic ties with Taiwan to recognise China are now receiving direct funding from China to help address the issue of climate change in their respective countries.

Conclusion

Climate change issues are a serious matter that affects the livelihood of the Pacific Islands countries. Both the acidity and temperature of seawater are inevitable, and mangrove ecosystems will be negatively impacted. Severe weather patterns associated with climate change have threatened the survival of mangroves. Powerful storms causing violent waves and tidal flow action will seriously impact mangroves that thrive in sheltered areas. Scientists have proven that man-made global warming is gradually producing conditions on Earth very similar to factors that preceded and caused natural catastrophic extinction events of plants and animals both in the oceans and on land in the past. These past events (meteorite impact and volcanic eruptions) increased atmospheric carbon and poisoned the oceans through acidification, which climate change is now doing.

With increased human activities within coastal areas in terms of human settlement, land use changes, the flow of solid and liquid waste, and coastal developments such as beach ramps, jetties, causeways, coastal protection structures, reef mining, and extraction of sand and beach aggregates, there is an ever-increasing change along Pacific coasts. In addition, climate change, climate variability, and extreme weather events have exacerbated the rate of change of Pacific coasts.

It is logically deducible that should the sea level increase on a more frequent basis, i.e., shifting from occurrences every few years to every few months, the island is likely to become essentially uninhabitable, characterised by a scarcity of freshwater resources and an inability to support crop cultivation. Given the existing trajectory of sea-level rise, this transformation could transpire within the next three decades or so (Storlazzi et al., 2015; Weir & Virani, 2011).

Pacific coasts are designated as important areas for providing vital Pacific livelihood. The coastal ecosystems, human settlement, and other major supporting services and basic infrastructure are centred on the coastal zone. Coasts are being used for many reasons. The underlying problem is that the Pacific coasts are in a state of crisis. A number of human engineering interventions over the past decades have contributed to and accelerated the coastal erosion problem in the Pacific. Pacific Island countries depend on marine resources for food and income. Given that most of the region's population is settled in coastal areas, changes in population density combined with new technology and changing development priorities have had a significant impact on coastal environments in the last decade.

Note

1 SOPAC miscellaneous publication no. 177 Coastal Protection in the Pacific islands: Current Trends and Future Prospects Proceedings of the First and Second Regional Coastal Protection Meetings: Held on 21–23 February 1994 in Apia, Western Samoa, and on 16–20 May 1994 in Suva, Fiji.

Bibliography

Ahmed, F., & Mishra, V. (2020). Estimating relative immediacy of water-related challenges in Small Island Developing States (SIDS) of the Pacific Ocean using AHP modeling. *Modeling Earth Systems and Environment, 6*, 201–214.

Cameron, C., Maharaj, A., Kennedy, B., Tuiwawa, S., Goldwater, N., Soapi, K., & Lovelock, C. E. (2021). Landcover change in mangroves of Fiji: Implications for climate change mitigation and adaptation in the Pacific. *Environmental Challenges*, *2*, 100018.

Kenchington, R. (2009). Maintaining coastal and lagoonal ecosystems and productivity. Proceedings of International Symposium onIslands and Oceans (pp. 1–11). Japan: Ocean Policy Research Foundation.

Lal, M. (2004). Implications of climate change in small Island developing countries of the South Pacific. *Fijian Studies: A Journal of Contemporary Fiji*, *2*(1), 15–35.

Marine Environment Education Program. (n.d.). *Primary school teachers guide.*

Mcleod, E., Bruton-Adams, M., Förster, J., Franco, C., Gaines, G., Gorong, B., James, R., Posing-Kulwaum, G., Tara, M., & Terk, E. (2019). Lessons from the Pacific Islands—Adapting to climate change by supporting social and ecological resilience. *Frontiers in Marine Science, 6*, 289.

Ministry of Economy. (2018). *Fiji low emission development strategy 2018–2050.* Government of Fiji. https://unfccc.int/sites/default/files/resource/Fiji_Low%20Emission%20Development%20%20Strategy%202018%20-%202050.pdf

Paeniu, L., Iese, V., Jacot Des Combes, H., De Ramon N'Yeurt, A., Korovulavula, I., Koroi, A., Sharma, P., Hobgood, N., Chung, K., & Devi, A. (2015). *Coastal protection: Best practices from the Pacific.* Pacific Centre for Environment and Sustainable Development (PaCE-SD), The University of the South Pacific.

Storlazzi, C. D., Elias, E. P., & Berkowitz, P. (2015). Many atolls may be uninhabitable within decades due to climate change. *Scientific Reports*, *5*(1), 14546.

Weir, T., Dovey, L., & Orcherton, D. (2017). Social and cultural issues raised by climate change in Pacific Island countries: An overview. *Regional Environmental Change*, *17*, 1017–1028.

Weir, T., & Virani, Z. (2011). Three linked risks for development in the Pacific Islands: climate change, disasters and conflict. *Climate and Development*, *3*(3), 193–208.

5 Climate Change as a Security Challenge

The Evolving Paradigm of International Regime for Climate Change in a Militarised World

Swasti Rao

Introduction

Climate change is one of the most tangible security challenges of our times. The effort to securitise the issue of climate change has sped up since the last decade owing to its increasing existential threat. Despite the growing realisation and development of international regime complex for climate change (Keohane & Victor, 2011, p. 8), the issue still suffers from what is called a *Climate Change Securitization Paradox* (Diez et al., 2016), where on the one hand are several securitisation moves by activists, government, and transgovernmental actors to rev up a national security debate around it, but on the other hand we have very rarely seen any extraordinary measures to tackle the threat of climate change.

Russia's war in Ukraine has upended the geopolitical and geoeconomic calculus of several countries. Its impact continues to cut across traditional and non-traditional security threats. The war has perpetuated a crisis of food, fuel, and fertilisers and pushed the world into a state of permacrisis.[1]

While most areas of collective action have been identified and acted upon to varying degrees by the global community, the disjunct between climate change enthusiasts and militaries has hardly been explored by mainstream environmentalists.

However, the impact of the return of conventional war and the resultant global militarisation[2] on climate mitigation and adaptation needs to be addressed with more urgency than ever before.

Over the years, burgeoning academic and scientific literature has unpacked the climate discourse and established climate change as a threat multiplier. However, even after years of research, climate adaptation and mitigation are still looked upon as non-traditional security challenges. Little academic engagement exists on the link between hardcore militaries and military alliances, which are some of the biggest polluters, and their efforts to set up regime complexes to deal with the challenge of climate change in the 21st century conflict scenarios. Until the war in Ukraine happened and Europe's security architecture was shredded to pieces, some green parties in Europe

DOI: 10.4324/9781003586869-6

and activists continued to oppose the very existence of armed forces. The two intuitively don't go together.

Approaching Climate Challenge in the World of Militaries: Regime Complex vs. Integrated Regime

While the dialogue around broader sustainability per se is not new, the scale and ambition of collective climate action are unprecedented. Climate change has long been known as a threat multiplier and is increasingly recognised as a 'shaping threat'[3] that fundamentally alters the environments in which militaries will have to operate in future decades. The subject of how militaries and alliances respond to this existential crisis then becomes a major imperative in any sustainable international regime complex on climate change. But it's not all a monolith and does not operate in a homogeneous domain of world action. In fact, quite the opposite. Keohane and Victor have argued that the structural and interest diversity inherent in contemporary world politics is such that despite repeated attempts, a comprehensive integrated regime on climate change continues to remain elusive. A decade later, with the growing ambivalence in the world order and the return of great power rivalry, the absence of an integrated approach has never been more palpable. On the contrary, the nature of the international system with conflicting interests and consequential hedging patterns makes it possible to generate a regime complex instead. A flexible regime complex allows for better management of the menace of climate change, especially when it comes to tackling the emissions of the world's largest militaries and military alliances.

Therefore, there is a need to *foreground* a more critical, multi-dimensional and a flexible approach that links the patterns of traditional militaries and military alliances and their role in the climate discourse. Part of the reason why these studies have not garnered enough traction is because, apart from a few advanced militaries, there is a lack of calculable and publishable data on military carbon footprints and emissions. Another problem is a lack of a multi-scalar approach that connects the various threads of traditional and non-traditional security aspects of the issue at hand across multiple scales and actors. When looked upon with a multi-scalar lens, military alliances' efforts to tackle climate change manifest real-world complex spatialities—economy, innovation, incentive, political, and military culture—that are usually not a part of the abiding climate discourse of our times. Once again, the distinction between regime complex and integrated regimes comes in handy.

The regime complex, in order to emerge effective, has to take a multi-scalar approach towards these complicated realities that are often at loggerheads. Integrated regimes with legally binding rules will never be able to permeate the space of norms around military emissions because the data itself is so scarce, partially reported, and national motivations are so mired in security dilemma.

Towards a Multi-scalar Approach of Climate Change Regime Complex in the Military Sphere

In the twenty-first century, almost all of the world's phenomena occur in open systems, which means that they are created by a variety of causal structures, mechanisms, processes, and fields (Bhaskar, 2010). The year 2020 was a paradigm shift in the way the world operated because of the pandemic, urging the whole world to take the issue of transboundary challenges in an interconnected world with more gravity and seriousness than ever. A similar paradigm shift has been prompted by the war in Ukraine and the subsequent global militarisation.

The chapter argues that a multi-scalar approach is better suited to explore the nuances and interconnectedness of militaries and military alliances and their fundamental reservations with the climate discourse. All the massive Greenhouse Gas (GHG) emissions inherent in the very actions of war, the vast and rampant joint military exercises, the testing of new weapons, and the constant pressure to test the interoperability of weapons and armed forces, notwithstanding the economic motive of the military-industrial complex that feeds on war technology and weapons—all these together present challenges that cannot be conceptualised without a multi-scalar analysis.

This chapter engages in a critical review of the contemporary climate discourse in the world's leading militaries and the largest military alliance, North Atlantic Treaty Organization (NATO), and the various efforts made by them to set up a climate change regime complex despite abiding limitations and backlash.

The chapter argues that NATO's example, although a work in progress, could serve as a template for our times for effective multilateral action around emission reduction and climate adaptation as conventional large scale war has made a comeback in the world.

The chapter concludes with the suggestion that including multi-scalar realities of a militarised world and subsequent efforts to tackle climate challenges of our times can qualitatively enrich the evolving regime complex for climate change, thereby substantiating deliberations on reducing the 'Climate Change Securitization Paradox'. The inclusion of this hitherto missing dimension is crucial in the evolution of the climate change regime complex at the international level for times to come.

Evolving Trends in Climate Discourse

The last five years have been important to the climate discourse because, for the first time, the world has seen a legally binding international treaty on climate change that for the first time has brought the major players together to negotiate and decide on Nationally Determined Contributions (NDCs), which have to be implemented and reviewed every five years. The Paris Agreement certainly has made the security regime more coherent and strengthened. Subsequent COPs have aided the process of review and progress, continuously trying to find convergences between arch-rivals on a matter that is intrinsic to the future of all humanity.

But when it comes to militaries and military alliances, there is little convergence in the template of practices around climate change and emissions. As highlighted in the previous sections, legally binding protocols are not likely to work in the sensitive domain of military emissions as they dissuade most militaries from developing apparatuses to quantify their military bootprint and then share it with other bodies. Accurate calculations and reporting of military emissions risk projections of military capabilities that a country at war or in border disputes may be wary of sharing with its adversaries. The advent of dual-use technology complicates the matter to another level. What is to be done?

First, defence needs to be decarbonised through multiple ways, creating economic incentives at the military-industrial level and multilateral standard setting. Integrated and binding regulations will only lead to defections.

Decarbonisning Defence in the Times of Global Militarisation

There is little doubt that the world today deals with an era of increasing military carbon bootprint. The first and foremost challenge at the global level is to reduce the greenhouse gas emissions caused by the militaries of the world.[4] It is rightly said that climate change leadership, especially in a militarised world, should seek to face hard truths rather than finding solace in stirring speeches.[5] What must also be remembered is that the major military actors that operate in the international climate change regime paradigm remain to be a handful of leading militaries in the world – most notably from the developed west that have reached a state of technological innovation that allows them in the first place to entertain sustainable innovations that enable reductions of emissions. Being long-standing democracies with institutional frameworks that more readily publish data on emissions is another major factor in innovating GHG reduction. Despite that, until recently the world's leading superpower, the United States, did not publish all the data on its military's greenhouse gas emissions. The situation with countries like China and Russia is intrinsically worse due to the type of ruling regimes.

Evolving Climate Acts of Big Militaries

The United States is not only is the most preeminent military power in the world, its Department of Defence is also the largest consumer of fossil fuel and the largest polluter globally. Analysts have remarked that the US alone is a bigger polluter and a climate actor than several of the world's industrialised countries by comparison.

Specialised studies aside, the path to calculating and publishing the consumption and emission patterns of militaries has been very arduous. One of the path-breaking achievements in the field, namely the Kyoto Protocol of 1997, had to concede compromise when it came to monitoring the military carbon footprint (bootprint) around the world. At the time, the compromise was seen as a middle ground in bringing the treaty to fruition. The US had arguments against sharing data for military emission monitoring to protect its unique position as the world's sole superpower. Treaties like the Kyoto Protocol, even though aimed at humanity's collective interest, were quite a victim of the unipolar world order (Lorincz, 2015).

The gap emanating from a failure to establish a working solution around one of the biggest sources of GHG, i.e., the militaries and particularly large militaries such as the US', continued well into the Paris Climate Agreement of 2015 (Neslen, 2015). Paris enthusiasts have been quick to point out the progress made since the Kyoto Protocol times. Paris seems a step ahead as, unlike previous protocols that never mentioned the issue, it gives individual countries the space to decide if they would like to reduce their military bootprint. However, in effect it hardly meant any development on the ground as militaries, by definition deal with a different nature of existential threats in today's complex world and assume a special entitlement to remaining disconnected from climatic clarion calls.

That is not all. Despite scientific research that has cautioned against massively growing military GHG emissions, it has largely remained an area protected by respective governments. Most governments, barring Western developed states, still do not have a comprehensive strategy to report their defence emissions.

Let alone taking steps to reduce emissions, most governments do not even have a national apparatus to calculate their annual military carbon emissions. This remains a structural impediment in the path of assessing the true magnitude of the problem at hand.

However, to their credit, for the first time, Crawford (2019) published a report by calculating data on US military energy consumption, which stood at nearly 60 million tons of carbon dioxide equivalent. According to the research, military bases and installations were responsible for 40 per cent of the emissions, and fuel use in military operations was responsible for 60 per cent.[6] Although this data, at best, is only roughly indicative of the mammoth problem at hand[7] it does give a useful insight into what a more militarised world of today needs to do to genuinely make a qualitative difference to develop a regime complex on climate change.[8]

Coming to Europe, the United Kingdom has a better track record of publishing data on its military carbon emissions than the US. The UK is one of the few countries in the world that have a national inventory of direct emissions from military aircraft and shipping (BEIS 2019). Moreover, the UK Ministry of Defence ranks first among government departments with the highest reported greenhouse gas emissions.[9] Therefore, setting an example, the country's total reduction targets include military aircraft and shipping as well.

It is no surprise then that the first call for tabulating military footprints at the Conference of Parties (COP) 26 came from Ben Wallace, the UK Secretary of Defence.[10] His is the only country in the world with comprehensive carbon emissions data across sectors, including the military, that is published annually by the government. Critics have argued[11] that even such approximations may not be perfectly representative, but they do offer a refreshing departure from the mainstay reservation elicited across primary actors on sharing their defence emissions.

Nonetheless, the UK has gone a step ahead and has sustainability outlined in the Climate Change and Sustainability Strategic Approach (CCSSA) published in March 2021.[12]

Recent Efforts by the UK and the US and Their Impact on Other Actors

There have been more recent efforts by the UK to link defence and climate change. Early this year, the UK's Defense Science and Technology Laboratory hosted its first climate change programme that focused on sustainability and climate security for defence. [13]

Following the UK's example, the US Army released its first-ever climate strategy in 2022 with three concrete goals.[14]

- reducing the US army's emissions to half by 2030
- *electrifying non-combat vehicles by 2035*
- *developing electric combat vehicles by 2050*

France, another European power inspired by the UK's example, has committed its armed forces to a three-pillar strategy of mapping climate risks, strengthening monitoring and research instruments, and supporting strategic research, designed to achieve climate neutrality by 2050.[15]

What about Other Parts in the World?

Data has been published by the Organization of Economic Cooperation and Development (OECD) on energy consumption and economic activity of their members, which may be used for estimating military carbon emissions as well. However, as noted earlier, transparency itself is a challenge with many governments, such as China, Russia, or Saudi Arabia. The problem is exacerbated as the developing countries of the world have systematically argued against the 'phasing out' approach to carbon while upholding the necessity of the 'phasing down' approach, as their economies remain more fossil-fuel dependent.[16] Needless to say, even in the absence of relevant data, larger carbon footprints can be inferred from across-the-board activities of developing economies.

When the above ambivalence is added to the travesty of war in general – ecological imbalance, scorched earth tactics, deforestation, mass migration, and the various dimensions of post-conflict rebuilding – the problem attains a mammoth proportion. Recognising the need to delve deeper into the world of military emissions in a very uncertain world, certain research methods have been devised lately. The next section discusses those.

Multi-scalar Methodological Pathways to Map and Reduce Military Emissions

Recognition of an essentially multi-scalar approach to developing a regime complex in climate change has enabled the development of corresponding methodological pathways. The developed militaries of the world have devised multi-scalar pathways to reduce the total pollution they cause. These range from efforts that are practical and consequential, such as improving fuel efficiency of military vehicles and using more renewable sources of energy, to more structural changes such as developing greener defence ecosystems. In the last few years, the results have been encouraging.

Making a transition to cleaner fuel, the UK's Royal Air Force (RAF) has recently successfully completed a Voyager air-to-air refuelling flight, powered by an approximately 43 per cent blend of Sustainable Aviation Fuel (SAF).[17] This significant achievement follows a series of recent milestones achieved by the RAF towards a sustainable aviation future, including the world's first RAF Voyager flight fuelled by 100% SAF.[18]

Are These Successes Enough?

Research shows that while the above successes are praiseworthy, they will do little to fundamentally cut global military emissions.

The most optimal strategy is to bring about reductions in military spending that will lead to necessary cuts in military pollution. The Intergovermantal Panel on Climate Change (IPCC) concluded that the only way to really reduce military carbon emissions is to reduce the huge military budgets around the world. However, the global trend is quite the opposite. There is no respite from militarisation, especially after the war in Ukraine and the looming crisis in Taiwan. Europe is militarising at an unprecedented pace, as are Japan and India. Russia's hydrocarbon sales are solely aimed at sustaining its war spending in Ukraine. Beyond Earth, outer space is getting militarised by the reckless and provocative activities of countries like China and the balancing measures taken by other spacefarers.

It is indeed a very depressing time to even entertain the utopian ideal of global demilitarisation.

The global community must brainstorm together. What are the other ways in which a country's defence sector can be made more climate-sensitive and sustainable?

Although difficult, incentives can be created so that the defence industries across the world find it viable to take such steps.

According to existing research, while defence spending cannot be curtailed, defence contractors could be given compelling business reasons to chase net zero.[19] Even though long-term, these can be summarised as follows:

- *Maintaining equity values and low capital costs*: As global investment trends towards becoming greener, climate change has been cited as influencing investment strategies for 86 per cent of fund managers surveyed by Robeco, a global asset manager, in its 2021 Global Climate Survey.[20] It is estimated that very soon, capital itself will undergo a green revolution. Defence contractors will be motivated to fall in line. While big defence manufacturers cannot become carbon neutral overnight, they can set targets to do so by 2050 in line with corresponding targets set by leading national militaries. Here, national governments have a major responsibility to ascertain that their strategic partnerships with each other and with defence manufacturers are modelled along sustainable practices.
- *Setting targets and National Defence re-adjustments*: Once again, US and UK defence departments have led the way forward when it comes to setting targets and following re-adjustments.

The UK Ministry of Defence had announced that it will start including environmental criteria in its tenders by 2022. In the US, the Biden Administration has committed to a 50 per cent national reduction of carbon emissions by 2030. To ensure this, the US has also launched a 'task force' to work out the implications for future defence contracts. It is a matter of time before the task force will start recommending changes to the Federal Acquisition Regulation, thereby institutionalising the change.

These changes are expected to bring awareness regarding the carbon efficiency of certain product and will make it easier to get them procured under new climate-friendly regulations. Clearly, these are templates to be followed by countries like India, which have made extensive defence acquisition procurement reforms to facilitate joint production ventures with foreign partners. With the rise of the push for competitive indigenisation of defence apparatus, ample space has emerged for the private sector to enter the domain. India, with a rising military stature and spending, can slowly and steadily create similar task forces to recommend green solutions in the acquisition and joint production of weapons. The chapter recommends that the recently signed Initiative on Critical and Emerging Technology (ICET) between India and the US[21] must, in the coming years, develop innovative solutions towards greener defence partnerships.

The role of R&D: But none of the above is likely to work until governments pledge to spend robustly on R&D. A greener defence ecosystem is a highly complex terrain that requires paramount technological innovation. The UK, for example, has pledged to more than triple R&D funding by 2024, to more than $30 billion, to 'build a future which is greener, safer, and healthier'. The US federal government also devotes a considerable portion of its $140 billion in annual R&D spending to sustainable technologies.[22] Private players like Airbus, for example, are investing in hydrogen R&D to design low-emission aircraft by 2035.[23]

Once again, this serves as a template for militaries around the world to increase the component of greener defence ecosystems.

Practical Solutions

Apart from the solutions for incentivising green defence production and acquisition that are inherently long-term, leading militaries have deployed practical solutions to reduce their emissions. The following illustrations underscore the value of practical solutions.

In the aerospace domain, reduced emissions can be achieved by designing more sustainable engines and engine fuels. In the defence sector, the US air force and many other leading actors have tested liquid hydrogen fuel cells for the same.[24] The US Navy's Great Green Fleet's program[25] has successfully shown the feasibility of greener fuels for naval vessels. The UK Ministry of Defence has worked on having a hybrid version of armoured vehicles that will be more environment friendly.[26]

As an impact of the above-incentivised and practical solutions, there has been a favourable change in the attitudes of the defence manufacturers as well. Most defence manufacturers have started to tackle the climate challenge in response to pressure from regulators, customers, investors, and the public. But at present, this development stands at a very rudimentary stage and needs a governmental push across the board to have any sizable impact on global military carbon emissions.

The Case of North Atlantic Treaty Organization (NATO)

At the multilateral level, NATO has been the first military alliance in history to come up with a set of strategies and action plans to reduce its carbon emissions at the ground level.

In June 2021, NATO leaders confirmed their shared political commitment to reduce defence emissions, labelling climate change, 'a defining challenge of our times'.[27] NATO Secretary-General Jens Stoltenberg is not only the first NATO leader to attend a UN climate change conference, he has also made climate change a pillar of the alliance.

NATO's climate plan provides a 360-degree approach and encompasses measures to increase both NATO's and its allies' awareness of the impact of climate change on security and defence. The noteworthy aspect of this set of measures is that they deliver emission reduction and climate mitigation without impacting the deterrent capabilities and defence posturing.

The action plan has four verticals for a comprehensive overhaul of the alliance's approach towards greener defence.[28] Closer analysis reveals the multi-scalar spirit espoused in the action plan.

- **Increased allied awareness:** NATO plans to conduct an annual Climate Change and Security Impact Assessment. This assessment will analyse the impact of climate change on NATO's strategic environment and its assets, installations, missions, and operations.
- **Adapt to climate change:** NATO will, building on its assessments, incorporate climate change considerations into its work on resilience, civil preparedness, defence planning, capability delivery, assets and installations, standards, innovation, training, exercises, and disaster response. NATO will also address the need to adapt its capabilities to the changing climate more prominently in its procurement practices and its partnership with industry.
- **Contribute to the mitigation of climate change:** The alliance will develop a NATO mapping and analytical methodology of greenhouse gas emissions from military activities and installations. This methodology will help allies' own emission assessment programmes and could contribute to formulating voluntary goals to reduce greenhouse gas emissions from the military.
- **Enhance outreach:** NATO looks to strengthen exchanges with partner countries, as well as with international and regional organisations that are active on climate change and security issues, including the EU, the UN, and others, where appropriate. NATO will also increase dialogue with civil society, academia, and industry on climate change and security issues to support its work and contribute to the global response to climate change.

In line with the above, the alliance has made significant progress by releasing the Secretary-General's Report on Climate Change and Security Impact Assessment in 2022.[29] NATO's Secretary-General Stoltenberg has announced the allies' decision to reduce gas emissions by at least 45 per cent by 2030, down to net zero by 2050.

NATO's Strategic Concept and Climate Change[30]: NATO's Strategic Concept of 2022 marks an inflection point of the alliance's evolution and its approach to broadening security concerns. Point 46 of the Strategic Concept underlines the following:

> The Alliance will lead efforts to assess the impact of climate change on defence and security and address those challenges. We will contribute to combatting climate change by reducing greenhouse gas emissions, improving energy efficiency, investing in the transition to clean energy sources and leveraging green technologies, while ensuring military effectiveness and a credible deterrence and defence posture.

This gets added relevance to countries in Asia, especially India, Japan, Australia, and South Korea as the Strategic Concept 2022 also marks the pivot of the alliance to the Indo-Pacific theatre. While, at this stage, it may be rudimentary to discuss the characteristics of this pivot, it shall suffice to underscore that, deducing from the latest Strategic Concept, climate change mitigation will remain a pronounced pitch of NATO's overtures in the region (Rao 2023). Therefore, NATO's capacity to play a more visible role in enriching the regime complex on climate change gets a further fillip in light of this development.

One of the most recurring themes across militaries' green defence engagement discussed above seems to have a fundamental dynamic.

The most obvious challenge for all militaries and/or military alliances is primarily one: how to reduce emissions while maintaining combat and deterrence capabilities? This brings us to another important multi-scalar methodological pathway, the Scope Model in defence emissions, developed to reduce GHG emissions of real-world operating militaries.

Scope Model for Reducing Global Military Emission

A detailed methodological framework has been constructed to tackle the most pressing concern for any military that is seeking to reduce its emissions. This framework is based on the concept of Scope emissions.[31]

Generally, Greenhouse Gas (GHG) emissions are divided into three categories for business organisations – *Scope-1, Scope-2, and Scope-3*. This system has been developed by the Greenhouse Gas Protocol. [32]

For reducing emissions, the organisation or the company will have to reduce emissions in all three categories. But why do these categories exist and what could be the difference in these?

To make sustainability by reducing emissions more realistic, scientific research on the subject has divided the emissions produced by militaries into three Scope categories.

These categories not only make the classification of the types of emissions easier, but they also create incentives for the defence forces to list their emissions by other trade-offs, thereby giving a way to better calibrate their emission reductions.

Scope-1 emissions: These are 'direct emissions' that operate from things that the company owns or controls.

Scope-2 emissions: These are 'indirect emissions' created by the production of the energy that a company or an organisation buys. An oft-quoted example is installing solar panels or sourcing renewable energy rather than using fossil-fuel-generated electricity that would cut Scope-2 emissions.

Applying this concept in the military sector, Scope-1 and Scope-2 are essentially those categories where defence forces can benefit from categorising their emissions for which they are directly responsible.

Scope-3 emissions: These are also 'indirect emissions' but they are different from Scope-2 as they refer to the supply chain itself that the organisation or a company cannot control directly. Usually, these form the biggest group of emissions.

Therefore, when applied in the military sector, Scope-3 emissions refer to that category of emissions that result from the supply chain, including direct or indirect suppliers.

Going further, drawing from the above classification, reduction measures in defence-related emissions can be classified as following:[33]

- ***Emissions that can be reduced***:Until an emission emanates from 'mission-critical' capabilities, it can be reduced with smaller measures. Most defence forces are likely to agree to emission reduction is feasible in this regard.
- ***Emissions that belong to mission-critical capabilities but can still be reduced***: These include reducing emissions from critical capabilities-orientated combat vehicles by electrifying their engines or making them more fuel-efficient.
- ***Emissions that belong to mission-critical capabilities that cannot be reduced***: These emissions, as a category are the hardest to change. For example, sorties requiring peak performance are made possible only through fossil fuel, whose performance will get marred if fuel is switched for fuel efficiency. These types of emissions fall under such a category.

These methodological pathways have emerged as a way of recognising practical and feasible ways to reduce carbon pollution without making impractical trade-offs. A defence manufacturer or a national government can assess its emission profile across the three scope emissions and bring necessary changes in each based on various assessments. Research shows that a reduction of up to about 60 per cent of Scope-1 and Scope-2 emissions is achievable by focussing on the first two types of defence emissions: those that are not linked to mission-critical capabilities and those that are linked to mission-critical capabilities but can be addressed without any impact on the mission (combat performance). All research reiterates that timely interventions can make a huge impact. Therefore, timing is key.

Summing up, considerable progress has been made from the days when most national governments did not publish their data on defence sector emissions. Research shows that Scope-1, Scope-2, and Scope-3 emissions can be reduced significantly if their emission sources are listed. The classification system itself has been a significant step forward in incentivising the defence sector on the ground. Future efforts in this direction should focus on reducing emissions from all possible scenarios, barring mission-critical emissions where peak performance is absolutely necessary.

In recent years, Europe has been a leader in turning to green defence as per the Scope Model, more specifically under Scope-1 and Scope-2. Research by Mckinsey[34] has shown positive trends emanating from Europe across industries that uphold the potential of renewable energy.

According to the report, examples available to defence leaders include the following:[35]

- ***Improving the efficiency of the defence estate and bases.*** This could be achieved with thermal insulation, systems that minimise water use, and improved waste-management systems.
- ***Transitioning to renewable heating.*** This includes use of biofuels or heat pumps powered by renewable electricity.
- ***Using renewable power sources.*** This includes solar systems that can take advantage of the large surfaces available (for example, rooftops or land on bases) to enable decarbonisation. Green energy can also be sold back into the grid.
- ***Installing carbon-capture technology***[36] *as* it becomes more efficient and affordable. Defence bases could be a logical site to install such technologies, particularly as defence forces build, operate, and occupy most of their estate. In that way, the benefit of any upfront costs that lead to future cost reduction will be fully captured.

As far as reducing emissions from Scope-3 are concerned, it is vital to identify the supply chain emissions along all the levels of the supply chain. In green military production, this refers to a cardinal part of resilient supply chains. One of the ways resilience can be promoted is by incentivising green goals and sustainable changes. The other way to ensure this is by creating a more favourable ecosystem that is conducive to more sustainable production and manufacturing, as discussed in the earlier sections.

Conclusion

The return of conventional war and global militarisation has once again brought humankind to the urgency of pushing forward a green revolution in military production and performance. The problem is a lack of a multi-scalar approach to link the military paradigm with conventional discourses on developing a regime complex on climate change. This is absolutely critical because, as argued in the chapter, integrated and binding regimes around controlling military emissions are more likely to lead to defections. Only a multi-scalar and flexible regime complex is best suited to address the challenge across all its dimensions.

The objective of the chapter has been to argue how to substantiate the regime complex on climate change through templates provided by advanced militaries and subsequent methodological pathways that can bring practical solutions to reduce military emissions. However, these innovations are limited to very few, although dominant, actors, and the vast majority of militaries are still not active in the co-development of this much-needed regime complex. The bigger challenge in the world today is to bring all military actors on the same page with regard to greener objectives. The developing world, such as India and the non-democratic world dominated by countries like China and Russia, are not even on board with agreeing to calculate their emission data let alone publish it. The reasons could vary, but the end product is a huge gap in ascertaining emission cuts by fast-developing militaries in the world.

What is required is urgent and sustained global attention directed towards linking global military emissions in a coherent system to incentivise emission cuts through three main channels elaborated by the Scope emission model.

The chapter argues that the UK, US, and France could lead the way for individual models of emission cuts while maintaining hard power projections. NATO could serve as a template of a military alliance with detailed vectors of engagement with the menace of climate mitigation and adaptation without compromising on combat readiness and peak performance. As it pivots to the Indo-Pacific, the developing countries of the region may be brought on board with better incentivising and exchange of views and opinions. Norms must be co-designed by all the stakeholders, and periodic reviews of developing countries' military emissions must become a shared goal.

Such measures are integral to developing an effective regime complex for climate change in a militarised world.

Notes

1 Permacrisis: what it means and why it's word of the year for 2022 (theconversation .com).
2 One Year After Russia's Invasion of Ukraine: Experts React | RAND.
3 The US Can't Compete With China Without Tackling Climate Change – Defence One.
4 How the world's militaries hide their huge carbon emissions (theconversation.c om).
5 Christian Cachola, How the world's militaries hide their huge carbon emissions (theconversation.com).
6 The same report argued that these were roughly equivalent to the total emissions of an industrialised country like Sweden or Switzerland.
7 Decarbonising defence carbon emissions | McKinsey.
8 Crawford (2019) calculated that the real figure would be at least four to five times the one made available for reference.
9 Ministry of Defence annual report and accounts 2019 to 2020 – GOV.UK (www .gov.uk).
10 Decarbonising defence carbon emissions | McKinsey.
11 The carbon bootprint of the military | SGR: Responsible Science.
12 Ministry of Defence Climate Change and Sustainability Strategic Approach (publishing.service.gov.uk).
13 Climate change conference addresses future needs in defence – GOV.UK (www.gov .uk).
14 US Army unveils climate strategy as military prepares for climate change – The Washington Post.

15 Presentation Climate and defence strategy.pdf (defense.gouv.fr).
16 Column: Coal trajectory is set whether it's 'phase out' or 'phase down': Russell | Reuters.
17 RAF Voyager refuels mid-air with Sustainable Aviation Fuel – GOV.UK (www.gov.uk).
18 Royal Air Force completes world-first sustainable fuel military transporter flight - GOV.UK (www.gov.uk).
19 The Growing Climate Stakes for the Defense Industry | BCG.
20 docu-202103-robeco-global-climate-survey.pdf
21 FACT SHEET: United States and India Elevate Strategic Partnership with the initiative on Critical and Emerging Technology (iCET) | The White House.
22 The Growing Climate Stakes for the Defense Industry | BCG.
23 ZEROe - Low carbon aviation – Airbus.
24 The epic attempts to power planes with hydrogen – BBC Future.
25 Great Green Fleet (navy.mil).
26 Armoured vehicles to test electric technology – GOV.UK (www.gov.uk).
27 NATO Climate Change and Security Action Plan 2021.
28 NATO - NATO Climate Change and Security Action Plan, June 14, 2021.
29 NATO - News: NATO releases its Climate Change and Security Impact Assessment, June 28, 2022.
30 290622-strategic-concept.pdf (nato.int).
31 What are Scope-3 emissions and how it differs from Scope-1 and Scope-2 | World Economic Forum (weforum.org).
32 Homepage | GHG Protocol.
33 For more details, refer to Decarbonising defense carbon emissions | McKinsey.
34 Europe's path to decarbonisation | McKinsey.
35 Europe's path to decarbonisation | McKinsey.
36 Driving CO_2 emissions to zero (and beyond) with carbon capture, use, and storage | McKinsey.

References

BEIS. (2019). Final UK greenhouse gas emissions national statistics: 1990–2017. https://www.gov.uk/government/statistics/final-uk-greenhouse-gas-emissions-national-statistics-1990-2017

Bhaskar, R. (2010). Contexts of interdisciplinarity: Interdisciplinarity and climate change. In R. Bhaskar, C. Frank, K. G. Høyer, P. Naess, & J. Parker (Eds.), *Interdisciplinarity and climate change. Transforming knowledge and practice for our global future* (pp. 1–24). New York: Routledge.

Crawford, N. (2019). Pentagon fuel use, climate change, and the costs of war. https://watson.brown.edu/costsofwar/papers/ClimateChangeandCostofWar

Diez, T., von Lucke, F., & Wellmann, Z. (2016). *The securitisation of climate change actors, processes and consequences*. London: Routledge.

Keohane, R., & Victor, D. (2011). The regime complex for climate change. *Perspectives on Politics*, 9(1), 7–23.

Lorincz, T. (2015). Messages for the Paris conference: The forgotten dimensions of climate change. https://www.sgr.org.uk/events/messages-paris-conference-forgotten-dimensions-climate-change

Neslen, A. (2015). Pentagon to lose emissions exemption under Paris climate deal. *The Guardian*. https://www.theguardian.com/environment/2015/dec/14/pentagon-to-lose-emissions-exemption-under-paris-climate-deal

Rao, S. (2023). NATO knocking on Indo-Pacific door. How India can make the best of this opportunity. *ThePrint*. https://theprint.in/opinion/nato-knocking-on-indo-pacific-door-how-india-can-make-the-best-of-this-opportunity/1608296/

6 Analysis of the Implementation of Hyogo Framework of Action (HFA)

From Local to Adaptive Disaster Governance and Resilience in Chiang Rai, Thailand

Maya Dania and Reni Juwitasari

Introduction

Chiang Rai is a province in Thailand's upper northern region, characterised by rugged topography. Chiang Rai is located on many active fault zones, including the Phayao, Mae Chan, and Mae Ing faults (Songlar et al., 2018). On May 5, 2014, at 6:08 pm, the biggest earthquake in the province was recorded with the earthquake's epicentre in Tambon Saikaow, Amphoe Phan, with a magnitude of 6.3 at a 7-km depth from the ground (Tanchaisawat & Hirano, 2018). While in Chiang Rai, the sudden earthquake was followed by building collapse, large surface cracks, liquefaction, injured people, and mortality in the population. The earthquake destroyed many structures in Northern Thailand's other provinces, including Mae Hong Son, Nan, Tak, Phrae, Phayao, Lampang, Lamphun, and Chiang Mai (Ratchadaporn, 2016).

The earthquake damaged many buildings in Northern Thailand's provinces, such as Chiang Mai, Phayao, Lamphun, Tak, Lampang, Nan, Phrae, and Mae Hong Son Province (Jintaprasat, 2016). After the quake, roads and bridges were unusable and public buildings such as temples, schools, and hospitals were heavily damaged (Suwanmolee, 2017). Chiang Rai is located about 785 kilometres from Bangkok and has over 1.2 million people living on 12,000 square kilometres of land divided into 18 districts, 122 sub-districts, and 1,655 villages. According to Thai statistical data (NSO, 2000), Phan District, as the near-epicentre, has a low population density with around 120,000 people working as farmers in the agricultural landscape. In Chiang Rai alone, the area within 30 kilometres from the epicentre was widely damaged and affected seven districts, including 50 sub-districts and 609 villages. It was reported that there were 107 injuries, one death, and 5,139 damaged houses.

According to the report from the Department of Mineral Resources, there were at least six geological fault lines in the North to be monitored after a 6.3-magnitude earthquake hit Chiang Rai on May 5, 2014, at around 6:08 pm. These fault lines are Mae Chan, Mae Ing, Phayao, Mae Tha, Thoeng, and Pua fault lines. Although generally, Chiang Rai has a reasonably low rate of seismic activities, the earthquake on May 5, 2014, produced nearly 1,000 aftershocks recorded by the regional and temporary seismic networks. Since the magnitude of occurrences and damages of rapid-onset natural

DOI: 10.4324/9781003586869-7

disasters, such as earthquakes, increased unprecedentedly in the northern region of Thailand, the local government in Chiang Rai Province needs to promote innovative and collaborative approaches to enhance effectiveness in disaster risk activities in parallel with the global commitment to Disaster Risk Reduction (DRR) as implemented by Thailand's Economic and Social Development plan.

The Hyogo Framework of Action (HFA) focusses on disaster risk reduction at the international, national, and sub-national levels to enable more methodical planning, execution, and assessment of disaster risk reduction initiatives. Additionally, it serves as a roadmap for building adaptive ability, boosting resilience, and decreasing disaster risk in 168 participating countries (including Thailand) via the integration of planning and policy, institutional development, and disaster risk reduction (Sternberg & Batbuyan, 2013; Djalante et al., 2012a). HFA has evolved into the primary worldwide tool for directing the implementation of DRR at all societal levels. The HFA was a 10-year action plan that lasted from 2005 to 2015, and disasters continued to inflict human, economic, infrastructure, and ecological losses across the globe, most notably in the world's most vulnerable and most impoverished nations. The HFA demonstrates a worldwide commitment to disaster risk reduction by establishing five action priorities as mentioned in UNISDR (2005):

1. Priority one: Ensure that DRR is both a national and a local priority, with a strong institutional basis for implementation
2. Priority two: Identity, assess and monitor disaster risks and enhance early warning
3. Priority three: Use knowledge, innovation, and education to build a culture of safety and resilience at all levels
4. Priority four: Reduce the underlying risk factors
5. Priority five: Strengthen disaster preparedness for effective response at all levels

Nonetheless, the HFA agenda expired in 2015, so promises to support DRR should be reaffirmed. As we enter the post-2015 era, the global commitment to disaster risk reduction is represented in the Sendai Framework for Disaster Risk Reduction (SFDRR), completed in March 2015 in Sendai, Japan. The recently adopted SFDRR (2015 – 2030) acknowledges that "Disasters, many of which are worsened by climate change and are rising in frequency and severity, substantially hinder progress toward sustainable development" (UNISDR, 2015) and therefore establishes four action priorities:

1. Priority one: Understanding disaster risk

Disaster risk management should be founded on a thorough knowledge of disaster risk in all of its aspects, including vulnerability, capability, exposure of people and assets, hazard characteristics, and the environment. Such information may be used to evaluate risks, avoid them, mitigate them, plan for them, and respond to them.

2. Priority two: Strengthening Disaster Risk Governance (DRG) to manage disaster risk

Disaster Risk Governance at the national, regional, and global levels is critical for catastrophe prevention, mitigation, preparation, response, recovery, and rehabilitation. It encourages cooperation and partnership.

3. Priority three: Investing in disaster risk reduction for resilience

Public and private investment in disaster risk prevention and reduction via structural and non-structural measures is critical for enhancing the economic, social, health, and cultural resilience of individuals, communities, nations, their assets, and the environment.

4. Priority four: Enhancing disaster preparedness for effective response and to "Build Back Better" in recovery, rehabilitation, and reconstruction

The increase in disaster risk necessitates improving disaster preparation for response, acting in advance of occurrences and guaranteeing efficient response and recovery capabilities at all levels. The recovery, rehabilitation, and rebuilding phases provide a significant opportunity to rebuild more effectively, particularly by incorporating disaster risk reduction into development efforts.

The chapter examines the DRG applied at the local government level in Chiang Rai through the lens of the 2014 Chiang Rai earthquake. The chapter illustrates how specific DRR policies have been successful, while others have persisted due to a lack of cooperation among DRG players. A theoretical discussion on governance under the DRR framework describes the HFA Priority 1 and the institutional capacity and DRR policies implemented at the local government level in Chiang Rai. Based on Ross' Local Disaster Resilience (2013) approach, it is revealed that the sub-national government organisation – the Sub-district Organization (SAO) in Chiang Rai – had made DRR a local priority and implemented DRG by Context (Disaster, Policy, and Development), Capacities (Institutional, Human Resources, and Financial), and Performance (Coordination). However, it is not robust since SAO has adaptive and resilient governance difficulties, as the latest terms proposed post-2015 are SFDRR.

Adaptive Government (AG) developed as a theoretical framework for governance as the appropriate solution to the systemic and complex issues associated with climate catastrophes resulting from SFDRR Priority two on improving Disaster Risk Governance (DRG). AG focusses on improving people's adaptive ability to cope with the complexity of catastrophes to create resilient institutions and social networks. In order to achieve sustainable development, catastrophe risk governance must be contextualised and linked to routes to resilience. As a result, it must be integrated into and affected by the broader governance of social, environmental, and technological change.

Conceptual Framework

This research developed a conceptual model of Integrated Disaster Management based on Ross' Local Disaster Resilience (2013) in order to comprehend the current DRG mechanism at the local level in Phan District, Chiang Rai Province, as well as a relationship model between Adaptive Governance and Resilience by Djalante et al. (2011) in order to analyse the future resilience goal that the SAO in Chiang Rai province must achieve.

Ashley D. Ross's (2013) have a conceptual framework on Local Disaster Resilience model and Djalante et al.'s proposed relationship model between Adaptive Governance and Resilience (2011). Local Disaster Resilience focusses on the ability of local governments to communicate the idea of resilience to emergency managers and local government officials. It is concerned with determining how resilience, as defined in HFA Priority one, is translated into disaster management strategies, especially at the local government level. Alternatively, the Adaptive Governance model aims to translate the SFDRR's prioritised activities and involve all stakeholders in disaster risk reduction to create disaster resilience. The Adaptive Governance model aims to engage all stakeholders in disaster risk reduction in order to achieve disaster resilience by interconnecting four primary characteristics:

- polycentric and multilayered institutions (multiple actors at multiple levels)
- participation and collaboration (leadership, trust, and social capital)
- self-organisation/networks (bridging organisation, boundary organisation, and transition arena), and
- learning and innovation (public/social learning and institutional memory).

Literature Review

The International Platform of Disaster Risk Reduction (DRR)

There is a global concern that natural catastrophes are growing more frequent, lethal, and costly. As a result, DRR has been designed to decrease disaster risks by examining and addressing the variables contributing to disasters, such as vulnerability reduction and enhanced preparation for unfortunate occurrences (UNISDR, 2012). Srikandini et al. (2018) assert that DRR platforms have grown prevalent in most disaster-prone nations due to global agenda commitment. Historically, since 1987, the United Nations (UN) has urged member states to establish national committees, or co-governance platforms, comprised of multiple actors engaged in disaster risk reduction activities under the auspices of the UN agency for the United Nations International Decade for Natural Disaster Reduction (UN-IDNDR). In 1994, the UN-IDNDR convened the First World Conference on Natural Catastrophe Reduction and developed the Yokohama Strategy and Plan of Action for a Safer World, effecting a paradigm shift away from a disaster-reaction culture toward one of disaster prevention. The UN-IDNDR was dissolved in 1999, and its effort to create a disaster-resilient community was continued by a successor organisation, the United Nations International Strategy for Disaster Reduction (UNISDR). In 2005, UNISDR convened a World Conference on Disaster Reduction and established the Hyogo Framework for Action (HFA) 2005–2015: Building Nations and Communities' Resilience as the internationally accepted framework for the complete and systematic implementation of disaster risk reduction initiatives (UNISDR, 2007; Djalante, 2012b). UNISDR (2007) defined *disaster risk reduction* as preventing new and reducing existing catastrophe risks to increase resilience.

Similarly, UNISDR (2008) recognises DRR as a conceptual framework for minimising vulnerabilities and disaster risks, as well as for avoiding (prevention) and mitigating (mitigation and preparation) the harmful effects of hazards. For this reason, HFA is conveyed in the structure of five priority actions in a newly proposed risk management pattern that begins with assessment, prevention, and mitigation, and concludes with preparedness and emergency response (Stanganelli, 2008). However, the DRR community has not consistently pursued the five priority activities in reality (Djalante et al., 2011). Thus, to achieve inclusive DRR at all community levels, it is essential to strengthen and modify how nations manage disasters at the sub-national level.

Disaster Risk Governance (DRG) and Resilience as a Global Agenda

Disaster Risk Governance (DRG) is frequently used in policy discourse and is linked to HFA Priority 1. DRG is a method of steering and governance that involves non-state players in the policy process via the use of coordination, cooperation, partnership, joint-working, alliance, collaboration, and networks to mitigate catastrophe risks and increase community resilience (Mardiah et al., 2017). DRG is concerned with disaster management (preparation, response, recovery, and rehabilitation) and seeks to understand the complex dynamics of institutional contexts, power relations, and policy advocacy in risk reduction (Srikandini et al., 2018). According to the UNDP (2015), DRG is founded on the principles of good governance such as participation, accountability, transparency, equality, effectiveness, and the quality of policymaking and public service delivery. Coles et al. (2004) believe that successful policymaking requires broad participation of the affected community. According to the study, catastrophe resilience will be ineffective without involvement with wider social groups. Thus, DRG practices should be implemented to increase the efficacy of DRR policies.

According to de la Poterie and Baudoin (2015), HFA Priority one refers to two urgent action plans on community involvement: (1) Recognise the significance and uniqueness of local risk patterns and trends; decentralise disaster risk reduction responsibilities and resources to appropriate sub-national or local authorities; and (2) Promote community participation in disaster risk reduction through the adoption of specific policies, the promotion of networking, and the strategic management of volunteer resources, as appropriate. As a result, adopting and executing local disaster risk reduction strategies and plans is considered one of the local government's responsibilities (Chatterjee et al., 2015). Additionally, UNISDR (2013) acknowledged nations' efforts in achieving HFA Priority one via establishing national policies, legal frameworks, and coordination structures.

Simultaneously, the Sendai Framework for Disaster Risk Reduction 2015–2030 (SFDRR) is the United Nations' first and current global policy framework for the post-2015 era, outlining a set of objectives and targets designed to include resilience in implementation activities and promote development (Aitsi-Selmi et al., 2015; Busayo et al., 2020; Wisner, 2020). It is a beacon of hope for integration and collaboration, predicated on the SFDRR making a difference where the HFA could not exist locally and methodically addressing underlying problems (Pearson and Pelling, 2015). The SFDRR's Priority three (Utilise knowledge, innovation, and education to foster a culture of safety and resilience at all levels) and Priority four (Reduce underlying risk factors) both have more references to resilience and disaster risk than the HFA does (Glantz, 2015). It moved the focus from disaster management to disaster risk management to foster a resilient culture (Etinay et al., 2018).

Methods

This chapter employs a qualitative approach to investigate the Disaster Risk Governance system enacted in Thailand and its implementation by local government entities in an exploratory case study. Qualitative research will gather data via a literature review and semi-structured field interviews, using empirical questions to examine a phenomenon in its naturalistic setting (Yin, 2012). The literature review is a critical stage in the research process of a qualitative study (Boote & Beile, 2005). Analysing the literature review entails dissecting a whole into its constituent pieces and reassembling them to determine the data collection's integrity. The data were gathered via literature study and semi-structured field interviews.

Primary data sources included scholarly journal articles, books, official reports issued by international organisations, and provincial and municipal development guidelines and plans. For example, the United Nations Development Programme (UNDP), the United Nations Office for Disaster

Risk Reduction (UNISDR), the 2005 National Civil Defense Plan, the 2010–2014 National Disaster Prevention and Mitigation Plan (NDPM), the 2005 National Preparedness Policy, the 2010–2019 Strategic National Action Plan (SNAP) for Disaster Risk Reduction, and the Ninth National Economic and Social Development Plan (2012–2016). Additionally, interviews with key stakeholders (local government officials and academic experts) were undertaken to compensate for data limitations and contextualise the literature study. The interview data were examined thematically to classify the main concepts corresponding to the study topics. The results of semi-structured interviews with key informants (local government representatives/SAO Phan District-Chiang Rai and experts) and an analysis of the DRG performance at the local government level in Chiang Rai are presented to contextualise the literature review.

The qualitative analysis research methodology is based on institutional analysis. Institutional analysis examines a complex system's governance model at various institutional levels and drivers. It aims to identify the drivers that contribute to the complexity of the disaster, the institutions that respond to the disaster's complexity, the policy instruments that are used, the effects of the instruments on actors, and actor perceptions of the effectiveness of the instruments in achieving their objectives (adaptive and resilience).

Results

According to UNISDR (2007), each country should support a National Platform that is government-owned and led to promote and build a complete national disaster risk reduction system that is country-specific. Thus, the HFA supports decentralisation to the sub-national level of provinces and districts, while the National Platform fosters upward into Regional and Global Platforms. The HFA is translated in Northern Thailand using a case study of a rapid-onset catastrophe after the 2014 Chiang Rai earthquake.

1) Context (Disaster, Policy, and Development)

Thailand's first establishment of a comprehensive national disaster risk reduction strategy was precipitated by the 2004 Indian Ocean earthquake and tsunami. At the time of the disaster, Thailand's only disaster-related laws were the 1979 Civil Defense Act (B.E. 2522) and the 1999 Fire Prevention and Suppression Act (B.E. 2542), with the Ministry of Interior establishing the Department of Disaster Prevention and Mitigation (DDPM) in 2002 as the primary government agency responsible for coordinating DRR activities. The body could not function properly since the new agency lacked legal foundations (Kitagawa, 2020).

Following the establishment of the HFA, Thailand formally approved the Disaster Prevention and Mitigation Act 2007, which was implemented on November 6, 2007, as specialised legislation authorising government entities at the national, provincial, and local levels to respond to natural catastrophes. The new Disaster Prevention and Mitigation Act 2007 is implemented as the National Law for disaster management administration, repealing the Civil Defence Act 1979 (B.E. 2522) and the Fire Prevention and Suppression Act 1999 (B.E. 2542), and establishing the Prime Minister-chaired National Disaster Prevention and Mitigation Committee. The National Disaster Prevention and Mitigation Plan 2010–2014 (B.E. 2553 2557) was a national guideline that outlined more effective disaster management methods under the country's Disaster Prevention and Mitigation Act 2007.

According to the structure outlined in the National Disaster Prevention and Mitigation Plan (NDPM 2010–2014), Thailand classifies disasters into four severity categories to designate the level of the responsible Director or Commander: small scale, medium scale, large scale, and large scale with catastrophic impact, as shown in Table 6.1. The 2010–2014 National Disaster Prevention and Mitigation Plan is divided into (1) Disaster Management Principles, (2) Disaster Response Procedures, and (3) Security Threat Management and Response Procedures. Additionally, the Disaster Prevention and Mitigation Act B.E. 2550 (2007) specifies and explains the catastrophe management arrangement, including disaster policy guidelines, operational procedures, and coordinating procedures.

(Source: National Disaster Prevention and Mitigation Plan (NDPM 2010–2014), Thailand)

The Ministry of the Interior created the Strategic National Action Plan (SNAP) on DRR based on the disaster management cycle in 2010 to prepare for and react to disaster scenarios between 2010 and 2019. SNAP establishes a disaster management plan consistent with the National Disaster Prevention and Mitigation Plan's fundamental principles to improve the protection of people and property. The plan identifies four elements as the SNAP's main strategies: (1) Preventative and Mitigative Actions, (2) Preparedness,

Table 6.1 Levels of a disaster severity scale and actors responsible

Level 1	*Small–scale disaster (Local Director, District Director, and Bangkok Metropolitan Director Assistant)*	*Containing the situation and suppressing the incident*
Level 2	Medium-scale disaster (Provincial Director and Bangkok Metropolitan Director)	Intervening on a broader authority scale
Level 3	Large-scale disaster with severe and widespread impact (Central Director and National Commander)	Intervening on a broader authority scale with the required specialist or specific equipment
Level 4	Large-scale disaster with catastrophic impact (The Prime Minister or Deputy Prime Minister)	Intervening in the most comprehensive authority scale, including integrated specialists or equipment

Table 6.2 SNAP's core strategies and activities in Thailand

Core strategy	*Examples of activities*
1. Prevention and Mitigation	Risk assessment, information management, Community-Based Disaster Risk Management Programs (CBDRM), and risk awareness initiatives
2. Preparation	Enhancing the early warning system, conducting catastrophe drills, developing a disaster preparedness strategy, preparing machines, preparing for basic requirements, allocating budgets, and preparing infrastructure
3. Emergency Response	surveillance, Incident Command System (ICS) management, evacuation planning, provision of humanitarian assistance, and search and rescue
4. Rehabilitation and Reconstruction plan	Assessment of catastrophe damage, measurement of relief activities, repair of urban infrastructure, distribution and administration of disaster relief items, mental health assistance, and development of a recovery plan

Source: (DDPM, 2010; Raungratanaamporn et.al., 2014).

(3) Emergency Response, and (4) Rehabilitation and Reconstruction Plan. Additionally, SNAP identifies the primary actors and sponsors for each strategic action, as detailed in Table 6.2.

2) Capacities (Institutional, Human Resources, and Financial)

DDPM has established operational processes and policies for managing DRR operations following HFA standards by adopting the SNAP launched by the government in cooperation with UNISDR (Fakhruddin & Chivakidakarn, 2014). According to the Deputy Director of the SAO (Sub-District Administration Office) in Phan District, Chiang Rai Province, when the earthquake struck on May 5, 2014, the shaking lasted about 40–45 seconds, followed by months of mild aftershocks. The earthquake was classified as having a hazard severity rating of two. Thus, the catastrophe administration was centralised on the Governor of Chiang Rai Province, who created an Emergency Operations Center (EOC) to command and manage the disaster-affected region under his jurisdiction. The Provincial Disaster Prevention and Mitigation Committee and the Bangkok Metropolitan Committee were established at the provincial level, tasked with disaster risk management by local government officials (Tanaka, 2009; Kitagawa, 2020). According to the Disaster Prevention and Mitigation Act 2007, DDPM is the primary national government agency responsible for disaster management. It is responsible for three types of disasters: (1) manufactured and natural, (2) disasters caused by air raids during wartime, and (3) disasters caused by sabotage or terrorist

Table 6.3 Governance mechanism for the 2014 Chiang Rai earthquake

Agencies	*Roles*
Provincial Administration Office (PAO) and Provincial EOC	Responded immediately to impacted residents in the affected region in the event of an emergency and dealt with donors and associated supporters
DDPM Chiang Rai	Connected Provincial EOC and SAO
Army disaster relief centre of Chiang Rai	Human resources, army hospitals, and heavy machinery were used to dismantle and destroy the damaged structure
Public health networks	Coordination of public health activities in Chiang Rai Provincial and North Regional Public Health Departments, as well as in other associated hospitals
Civil society volunteers, academic institutes, NGOs, and the coordination centre of building inspection	Supported the Provincial EOC and PAO, as well as other supporting networks
Local Administration Office (LAO) with Sub-district Administration Office (SAO)	Provided information on the impacted region, loss, and casualties in order to facilitate the distribution of emergency assistance

Sources: Provincial Disaster Prevention and Mitigation Committee and the Bangkok Metropolitan Committee, Interviews with the head of sub-national leaders in Chiang Rai Province.

attacks – Table 6.3 details the disaster management in Chiang Rai after the 2014 earthquake.

3) Performance (Disaster Risk Reduction Coordination)

After the primary shock, the provincial administration (PAO – Provincial EOC) immediately declared catastrophic Stage 3 as specified in SNAP for all local actors, as described in this Table 6.4.

Source: Interview with the Deputy Director of SAO, Phan District, Chiang Rai Province, 2020; Suwanmolee, 2016).

According to the interview and report, during the Emergency Response phase's first 24 hours, the local government recognised the mechanism for providing basic needs assistance to people in systematic arrangements: food, drinking water, clothing, health, and rescue, as well as collecting initial situation data. The local government offered shelter, properties, amenities, and Search and Rescue (SAR) for missing people or corpses for the following 24–48 hours. Within 48–72 hours, the local government created a coordination centre for building inspections and performed loss assessment, preliminary relief, and long-term recovery after 72 hours, including impact assessment and victim rehabilitation. Despite this, the recovery phase was postponed, and instances of duplicate compensation reimbursement were discovered.

Table 6.4 Response actions after the 2014 earthquake in Chiang Rai Province

Declaration phases	*Duration*	*Agencies and actions*
Emergency Response	1 hour (Sudden interruption caused by a blackout and a breakdown of telephone/ internet connections as a result of the shaking)	• Village authorities directed residents to exit their buildings and go to an outdoor location. • The province governor created an Emergency Operations Center (EOC) at Chiang Rai city hall to deal with the earthquake. The hospital in the impacted region moved patients from a high-rise to the ground and then established a three-day field hospital.
	2 hours later	• Chiang Rai's Army disaster relief centre, Disaster Preparedness and Mitigation Region 15 (DPMRC15), and Highway police reacted to the Chiang Mai-Chiang Rai road collapse.
	24 hours later	• The Chiang Rai Army Disaster Relief Centre and the Disaster Preparedness and Mitigation Region Chiang Rai 15 (DPMRC15) created a Front Operation Centre in the impacted region.
	48 hours later	• The Chiang Rai Public Works and Town & Country Planning offices created a building inspection coordination centre.
	72 hours later	• The DDPM Chiang Rai, public works agencies, academic institutions, and provincial administration examined and maintained the data on the damaged buildings.
Recovery Stage	Within four days	• Chiang Rai authorities developed a civic society for disaster management.
	Within 14 days	• The North Regional Public Health Department has established a mobile primary care and mental health unit.
	Within one month	• The local government hosted a workshop and training session for local construction contractors on repairing and reinforcing residential structures damaged by earthquakes.

(*Continued*)

Table 6.4 (Continued)

Declaration phases	*Duration*	*Agencies and actions*
	Three months after	• Chiang Rai civil society for disaster management organised a public seminar to share lessons learnt in earthquake recovery.
Preparedness Process	Nine months after	• The DPMRC15 and the DDPM Chiang Rai have established the One School, One Search, and Rescue (OSOS) Program. The Provincial Disaster Prevention and Mitigation Plan was practised by DPMRC15, DDPM Chiang Rai, Chiang Rai Army, and associated agencies.
	One year after	• DDPM Chiang Rai enhanced the Disaster Prevention and Mitigation Plan at the sub-district, district, and provincial levels.

Additionally, the SAO officials admit in the interview that Chiang Rai Province has historically been overwhelmed by calamities like floods, rainstorms, and periodic droughts. However, past seismic devastations were not expected. As a result, an alert system and emergency management were ineffective in dealing with moderate-to-large-scale impact catastrophes. Typically, preparations for dealing with rapid-onset disasters in the province focus on the features of catastrophes within the scope of severity at a given stage. While this is true, Chiang Rai faces future earthquakes due to its geographical position between active earthquake zones.

Discussion

The results indicated that the SAO in Chiang Rai had adopted DRG by Context (Disaster, Policy, and Development), Capacity (Institutional, Human Resources, and Financial), and Performance (Disaster Risk Reduction Coordination) in line with HFA Priority one implementation at the local level. However, it is not entirely efficacious since SAO continues to have difficulties with Adaptive Governance, namely polycentrism, cooperation, self-organisation, and creativity. Disasters have become more complicated over the last decade, with vulnerable people and the environment intertwined. According to Djalante et al. (2011), disaster risk reduction (DRR) is a systematic method for managing catastrophe risks, while Adaptive Governance (AG) is proposed as a complementary approach to regulating complex issues such as disasters.

Many obstacles to converting SFDRR into AG have been identified on the ground. These include a lack of political will, the battle to collect adequate resources to support DRR at various levels, and different local capabilities affecting community involvement and decentralisation (Manyena et al., 2013; Jones et al., 2014). As Djalante et al. (2011) noted, AG necessitates developing new governance structures that are less rigid, more uniform, less prescriptive, and less hierarchical, offering a more creative yet successful approach to complex environmental issues. Additionally, Djalante et al. (2012a) discovered that one of the obstacles to enhancing DRR at the local level is a lack of financial resources to support risk mitigation and recovery efforts. Thus, greater cooperation between local governments and communities would aid disaster risk reduction planning and assist in developing future frameworks for reducing catastrophe risks (Mercer et al., 2008; Kelman and Glantz et al., 2014; de la Poterie and Baudoin, 2015).

1. Polycentric and multilayered institutions (multiple actors at multiple levels)

During the emergency for hazard severity level two, the provincial government (PAO), provincial EOC, DDPM, and Army disaster relief centre were deemed the most dependable players in quickly avoiding social turmoil. The rapid-onset disaster in Chiang Rai in 2014 demonstrated that the relationships between each agency, including the PAO, EOC, the DDPM Chiang Rai, Army disaster relief centre, public health networks, civil society volunteers, academic institutes, non-governmental organisations (NGOs), and the coordination centre for building inspection, were all independent, as they operate independently. According to their functional responsibilities, each agency had its database system and expertise in DRR regulations. As a result, disaster management continues to be a matter of hierarchy (vertical relationship) and rigorous adherence to standard operating procedures based on Acts or regulations enacted in disaster management initiatives.

2. Participation and collaboration (leadership, trust, social capital)

According to the 2007 Disaster Prevention and Mitigation Act, all Ministries and Departments are accountable for disaster recovery aid in some capacity or another, based on their capabilities and commitment. Nonetheless, not all administrative levels have the resources necessary to execute long-term DRR policies. According to the LAO representative in Phan District, current rules are adequate to a certain degree since the local governing body would immediately offer aid, repair, and assess the disaster-related damage.

3. Self-organisation/networks (bridging organisation, boundary organisation, transition arena)

According to SNAP, Thailand has designated vital players and supporters to coordinate national policy at the provincial and local government levels. Most of the actions and initiatives outlined in SNAP are organised by DDPM as the primary actor, with significant assistance from PAO, LAO, and SAO. According to the DDPM, Thailand's disaster scale classifies the 2014 Chiang Rai earthquake as a medium-scale disaster, necessitating immediate action by the provincial governor to control the situation with vertical coordination to the district, sub-district, and village levels of the response action.

Even though Chiang Rai has a robust DRG framework for the emergency phase, the long-term recovery plan is often less successful due to insufficient financial resources for DRR operations. In Chiang Rai, the preliminary plan for the local development plan often places a higher premium on infrastructure development than on disaster prevention and mitigation. In reality, the local authority at SAO lacks appropriately qualified human resources and the capacity to deal with disasters, particularly in victim aid missions. Local administration offices at SAO in impacted districts played a critical role in assessing and collecting data from their permitted area and submitting the report to the district's disaster administration board for approval before forwarding it to the office of DDPM Chiang Rai. The report mandated that village leaders and SAO committees promptly conduct extensive damage and loss assessments. The study was subsequently utilised as the basis for budget/resource distribution and disaster compensation reimbursement among LAO in Chiang Rai Province's impacted districts. Additionally, redundancy in compensation reimbursement was discovered because SAO was required to perform all data collection, verification, and distribution to the DDPM office and EOC, despite limited information systems, human resources, budget, and coordination in sharing database systems and knowledge transfer.

4. Learning and innovation (public/social learning and institutional memory)

Clearly, when it came to controlling DRR operations in Chiang Rai, the local government continued to use conventional disaster management techniques, including bureaucratic command and control, during an emergency. At the same time, the SAO's activity relied on the EOC as the central network, the DDPM in Chiang Rai, and the Army disaster relief centre as the emergency phase's coordinator for human resources. Similarly, DRR has not been elevated to a high priority in provincial development plans, although provincial governments are urged to create DRR activities and budgets for plan implementation and exercise. The sole budgetary resources come from the central government, which provides funds to assist plan creation for the provincial government at least once a year to guarantee the plan's efficacy and application. As a result, command-and-control methods remain a top-down approach (vertical direction).

Conclusion

HFA has evolved into a critical component of the worldwide instrument for implementing DRR at all levels of society. Thailand has ratified the HFA and incorporated international principles into the country's national laws, as shown in the Disaster Prevention and Mitigation Act 2007. With the 2014 Chiang Rai earthquake serving as a case study, the implementation of HFA Priority one in rural Northern Thailand demonstrates a lesson for the DRG in a developing nation. However, there was still a barrier to transforming the DRG mechanism into an adaptive and resilient government at the local level in Northern Thailand in order to achieve the current post-2015 SFDRR priorities, particularly in achieving SFDRR Priority two in strengthening the DRG into polycentric and multilayered institutions, participation and collaboration, self-organisation/networks, and learning and innovation, particularly in Northern Thailand.

This study makes recommendations, particularly to local governments that serve as the primary suppliers of different community services and implementing agencies for most DRR initiatives, using transformational leadership in controlling catastrophe risks. According to the report and studies based on the 2014 Chiang Rai earthquake case study, it is preferable to change the DRG toward adaptive and resilient government at the local level from vertical to horizontal directions in order to reduce command and control at the top and increase the government's flexibility and adaptability at the local level. The ability of the government at the local level to self-organise in terms of information systems, human resources, budgeting, and coordination in terms of sharing database systems and knowledge transfer would be enhanced. Local governments must play a significant role in local disaster risk reduction. Thus, further development and implementation of SFDRR through the post-2015 development and implementation of local laws, regulations, and public policies that include all communities are critical points for strengthening Disaster Risk Governance and reducing disasters.

Acknowledgement

The chapter received funding support from Mae Fah Luang University Research Fund (2019–2020), Mae Fah Luang University, Chiang Rai, Thailand.

References

Aitsi-Selmi, A., Egawa, S., Sasaki, H., Wannous, C., & Murray, V. (2015). The Sendai framework for disaster risk reduction: Renewing the global commitment to people's resilience, health, and well-being. *International Journal of Disaster Risk Science*, 6(2), 164–176.

Boote, D. N., & Beile, P. (2005). Scholars before researchers: The centrality of the dissertation literature review in research preparation. *Educational Researcher, 34*(6), 3–15.

Busayo, E. T., Kalumba, A. M., Afuye, G. A., Ekundayo, O. Y., & Orimoloye, I. R. (2020). Assessment of the Sendai framework for disaster risk reduction studies since 2015. *International Journal of Disaster Risk Reduction, 50*, 101906.

Chatterjee, R., Shiwaku, K., Gupta, R. D., Nakano, G., & Shaw, R. (2015). Bangkok to Sendai and beyond: Implications for disaster risk reduction in Asia. *International Journal of Disaster Risk Science, 6*(2), 177–188.

Coles, E., & Buckle, P. (2004). Developing community resilience as a foundation for effective disaster recovery. *The Australian Journal of Emergency Management, 19*(4), 6–15.

de la Poterie, A. T., & Baudoin, M. A. (2015). From Yokohama to Sendai: Approaches to participation in international disaster risk reduction frameworks. *International Journal of Disaster Risk Science, 6*(2), 128–139.

Djalante, R. (2012). Adaptive governance and resilience: The role of multi-stakeholder platforms in disaster risk reduction. *Natural Hazards and Earth System Sciences, 12*(9), 2923–2942.

Djalante, R., Holley, C., & Thomalla, F. (2011). Adaptive governance and managing resilience to natural hazards. *International Journal of Disaster Risk Science, 2*(4), 1–14.

Djalante, R., Thomalla, F., Sinapoy, M. S., & Carnegie, M. (2012). *Building resilience to natural hazards in Indonesia: Progress and challenges in implementing the Hyogo Framework for Action.* https://www.researchgate.net/publication/257632943_Building_resilience_to_natural_hazards_in_Indonesia_Progress_and_challenges_in_implementing_the_Hyogo_Framework_for_Action

Etinay, N., Egbu, C., & Murray, V. (2018). Building urban resilience for disaster risk management and disaster risk reduction. *Procedia Engineering, 212*, 575–582.

Fakhruddin, S. H. M., & Chivakidakarn, Y. (2014). A case study for early warning and disaster management in Thailand. *International Journal of Disaster Risk Reduction, 9*, 159–180.

Glantz, M. H. (2015). The letter and the spirit of the Sendai Framework for Disaster Risk Reduction (aka HFA2). *International Journal of Disaster Risk Science*, 6(2), 205–206.

Jintaprasat, R. (2016). *Soil amplification assessment of earthquake ground motion using geophysical and geotechnical data in Amphoe Muang, Changwat Chiang Rai* [Doctoral thesis]. Faculty of Science, Chulalongkorn University.

Jones, S., Oven, K. J., Manyena, B., & Aryal, K. (2014). Governance struggles and policy processes in disaster risk reduction: A case study from Nepal. *Geoforum, 57*, 78–90.

Kelman, I., & Glantz, M. H. (2014). Early warning systems defined. In *Reducing disaster: Early warning systems for climate change* (pp. 89–108). Springer.

Kitagawa, K. (2020). Development of disaster risk reduction policy in Thailand. *Disaster Prevention and Management: An International Journal.* https://discovery.ucl.ac.uk/id/eprint/10093407/1/Kitagawa_PDF_Proof.pdf

Manyena, S. B., Mavhura, E., Muzenda, C., & Mabaso, E. (2013). Disaster risk reduction legislations: Is there a move from events to processes?. *Global Environmental Change, 23*(6), 1786–1794.

Mardiah, A. N., Lovett, J. C., & Evanty, N. (2017). Toward integrated and inclusive disaster risk reduction in Indonesia: Review of regulatory frameworks and institutional networks. In Djalante, R., Garschagen, M., Thomalla, F., & Shaw, R. (eds) *Disaster Risk Reduction in Indonesia* (pp. 57–84). Springer, Cham. https://doi.org/10.1007/978-3-319-54466-3_3

Mercer, J., Kelman, I., Lloyd, K., & Suchet-Pearson, S. (2008). Reflections on the use of participatory research for disaster risk reduction. *Area, 40*(2), 172–183.

National Statistical Office of Thailand (NSO). (2000). *Population statistics*. http://web.nso.go.th/
Pearson, L., & Pelling, M. (2015). The UN Sendai framework for disaster risk reduction 2015–2030: Negotiation process and prospects for science and practice. *Journal of Extreme Events*, 2(1), 1571001.
Raungratanaamporn, I. S., Pakdeeburee, P., Kamiko, A., & Denpaiboon, C. (2014). Government-communities collaboration in disaster management activity: Investigation in the current flood disaster management policy in Thailand. *Procedia Environmental Sciences, 20*, 658–667.
Ross, A. D. (2013). *Local disaster resilience: Administrative and political perspectives*. Routledge.
Songlar, T., Pussadee La-or, N. P., Chomchoe, C., & Khunthason, S. (2019). Knowledge, attitude and practice (KAP) of earthquake preparedness amongst the elderly in risk areas: Chiang Rai, Thailand. *Journal of Health Research*, *33*(1), 2–13.
Srikandini, A. G., et al. (2018). Disaster risk governance in Indonesia and Myanmar: The practice of co-governance. *Politics and Governance,* 6(3), 180–189.
Stanganelli M. (2008). A new pattern of risk management: The Hyogo framework for action and the Italian practice. *Socio-Economic Planning Sciences,* 42(2), 92–111. http://dx.doi.org/10.1016/j.seps.2006.10.001
Sternberg, T., & Batbuyan, B. (2013). Integrating the Hyogo Framework into Mongolia's disaster risk reduction (DRR) policy and management. *International Journal of Disaster Risk Reduction, 5*, 1–9.
Suwanmolee, S. (2017). *Social* network analysis *of* disaster response *in 2014 Chiangrai province earthquake* [Conference session]. International Conference on Disaster Management: From Polar Region to the Local Communities Social and Environmental Development National Institute of Development Administration (NIDA). https://www.researchgate.net/publication/316134398_Social_Network_Analysis_of_Disaster_Response_in_2014_Chiangrai_Province_Earthquake
Tanaka, N. (2009). Vegetation bioshields for tsunami mitigation: A review of effectiveness, limitations, construction, and sustainable management. *Landscape and Ecological Engineering, 5*(1), 71–79.
Tanchaisawat, T., & Hirano, N. (2018). Appropriate attenuation model for Chiang Mai, Thailand from field measurement to model equation. *International Journal of GEOMATE, 15*(47), 48–52.
The National Disaster Prevention and Mitigation Plan (NDPM). (2010–2014). The Department of Disaster Prevention and Mitigation (DDPM), Ministry of Interior, Thailand [in Thai].
The Ninth National Economic and Social Development Plan. (2002–2006). The Office of National Economic and Social Development Board (NESDB), the Prime Minister's Office, Thailand [in Thai].
The Strategic National Action Plan (SNAP) for Disaster Risk Reduction. (2010–2019). The Department of Disaster Prevention and Mitigation (DDPM), Ministry of Interior, Thailand [in Thai].
UNDP. (2015). *National post-disaster recovery planning and coordination.*
UNISDR (United Nations International Strategy for Disaster Reduction). (2005). *Hyogo framework for action 2005–2015: Building the resilience of nations and communities to disasters.* http://www.unisdr.org/files/1037_hyogoframeworkforactionenglish.pdf
UNISDR. (2007). *Hyogo framework for action 2005–2010: Building the resilience of nations and communities to disasters.* http://www.unisdr.org/2005/wcdr/intergover/official-doc/L-docs/Hyogo-framework-for-action-english.pdf.
UNISDR. (2008). *Briefing note 01: Climate change and disaster risk reduction.*

UNISDR. (2012). *Disaster risk and resilience*. Thematic Think Piece, UN System Task Force on the Post-2015 UN Development Agenda.
UNISDR. (2013). *Disaster Risk Reduction in the United Nations Roles, Mandates and Results of Key UN Entities*. Available at: https://www.unisdr.org/files/32918_drrintheun2013.pdf
UNISDR. (2015). *Sendai framework for disaster risk reduction 2015–2030*.
Wisner, B. (2020). Five years beyond Sendai—Can we get beyond frameworks? *International Journal of Disaster Risk Science*, *11*(2), 239–249.
Yin, R. K. (2012). *Application of case study research* (3rd ed.). Sage.

7 Climate Change in the Indian Ocean Islands

Adaptation and Mitigation Strategies

Mohd. Faheem

Introduction

Climate change is responsible for influencing and shaping natural ecosystems and habitats. It leads to disruptions in natural food chains and also causes water and vector-borne diseases. The world is concerned about the changes occurring at present because they are accelerating and significantly impacting people's lives. As a result, action is being taken as quickly as possible to try to remedy this issue or stop more severe climate change from happening. The International Panel on Climate Change (IPCC) experts estimate that by the year 2100, the sea levels might rise by seven to 23 inches because of global warming and the melting of glaciers. The Small Island Developing States (SIDS) continue to face such challenges and may be submerged under the sea in times to come. This paper aims to analyse the impacts of climate change in the Indian Ocean, particularly on small developing island countries, and also find solutions in terms of adaptation and mitigation strategies to tackle the impact of climate change in the region of the Indian Ocean.

Climate Change in Indian Ocean

Globally, environmental pressures on oceans and coastal areas are increasing. In addition to habitat loss and biodiversity loss, land- and marine-based pollution, resource overuse, and the expanding effects of greenhouse gas emissions and global climate change are some of these growing pressures. No part of the ocean is immune to these challenges, and 41 per cent of the world's seas experience several stresses. The Indian Ocean Region (IOR) littoral countries are among the most exposed to environmental concerns associated with coastal living, with significant areas of the Indian Ocean currently ranking among the planet's most severely damaged marine ecosystems.

Indian Ocean observations are made from both ground and satellite sources. Since the 1960s, a distinct regional pattern in sea-level rise has been discovered using climate change simulation models of the ocean. In some sections of the Indian Ocean, sea levels have risen, but in southern tropical regions, they have significantly declined, according to the experts. The third-largest ocean in the world, the Indian Ocean contains around 20 per cent

DOI: 10.4324/9781003586869-8

of the planet's surface water. There is a vast area of warm water within this ocean that stretches from the east coast of Africa to the Pacific Date Line. This warm pool in the Indo-Pacific has warmed by around 0.5°C during the past 50 years. Sea levels have increased by 12.7 mm on average every 10 years along the northern coast of the Indian Ocean. Particularly affected regions include the Bay of Bengal, the Arabian Sea, Sri Lanka, Sumatra, and Java. These modifications to localised ocean circulation and sea-level rise can potentially worsen monsoon-related flooding in countries such as Bangladesh and India. Also, it can lead to irregular precipitation patterns. Islands in the middle of the Indian Ocean may see much larger sea level rises than the world average due to climate change effects on the Indo-Pacific warm pool. As a result, several Indian Ocean beaches and islands, as well as their inhabitants, will experience more environmental stress.

Another study on sea-level rise (Webb & Kench, 2010) seems to refute the widespread belief that low-lying reef islands will eventually become uninhabitable due to the effects of climate change. It compares the landform dynamics of 27 atoll islands in the central Pacific Ocean using aerial and satellite photographs collected during the previous 60 years, a time period during which there is evidence that sea levels have risen. According to the study, the islands' overall land area increased by 63 hectares (7 per cent) rather than decreased. The study's findings demonstrate that while all 27 islands underwent physical change and sea levels in the central Pacific Ocean rose by about 2.0 mm/year, there was significant variation in the pace and type of change among the islands, leading to a net increase in land area; 86 per cent of the islands either remained relatively stable or grew in size. In only four of the 27 islands was there area reduction of more than 3 per cent, whereas 12 of the 27 islands, there was an increase in size of more than 3 per cent.

The islands' locations on the reef platform have also shifted. For instance, the islands shifted toward their lagoons in 65 per cent of the cases. Islands are a persistent component of atoll reefs and can change in size despite sea-level rise. The island boundaries may adjust in response to a variety of variables. Finally, while thinking about how to adapt to the effects of a changing climate, the islands must consider their own characteristics and the rate of change.

One of the biggest risks to people's lives and means of subsistence in coastal areas worldwide is climate change. It will dramatically increase already existing risks, such as flooding from storm surges and cyclones. Other climate-related concerns, such as drought, saline intrusion, sea-level rise, and changes in temperature and rainfall, are increasingly posing severe risks to humankind's security of food, water, energy, and health. Climate change poses a serious threat to the coastal nations of the Indian Ocean Region, which stretches from East Africa to Southeast Asia. Many are particularly under peril due to their location and geography. Because a huge amount of their population resides in significant river deltas that are vulnerable to sea-level rise, countries like Bangladesh and Vietnam, for instance, are anticipated to be among the most severely affected in the world (Dasgupta, 2007). Cyclones and storm surges have a greater impact on Bangladesh's low-lying coastal plains because of the narrow and funnel-shaped Bay of Bengal. About 29 tropical cyclones struck Bangladesh in the second half of the twentieth century, with one of the deadliest killing 138,000 people in April 1991 (Haider et al., 1991).

People are at high or extremely high risk of having their homes damaged by flooding in several parts of the Indian Ocean. However, the Indian Ocean Islands, where one-third of the population believes they are at a high or very high danger of losing crops or cattle, tend to be more concerned about the issue of agricultural loss due to flooding. The sensation of loss from flooding is matched by this enhanced perception of vulnerability.

Socio-economic Impacts of Climate Change in the Indian Ocean Islands

The Indian Ocean Region (IOR), home to crucial ecosystems, is experiencing climatic change. More than 70 million square kilometres of land are covered by the Indian Ocean. The 38 nations that surround it are home to roughly one-third of the world's population. The fisheries sector is still arguably the most significant because it is a major source of food and a means of subsistence for coastal communities, even though their economies' so-called blue economy encompasses a number of important economic sectors, including shipping, tourism, minerals, oil and gas, maritime transport, etc. However, because of climate change, the sector is currently dealing with severe difficulties. This includes changing ocean chemistry and warming ocean waters. Over the past 20 years, the heat content of the world's oceans has increased by more than 30 per cent, with the Indian Ocean in particular warming at an alarming rate. Distribution, abundance, populations, and variety of fish have all been impacted by this. In addition, additional human-caused stresses including habitat loss, overfishing, and marine pollution interact with climate change. The fishery resources of the Indian Ocean are therefore in grave danger, as are the millions of people who depend on them for their sustenance and way of life.

Globally, climate change will have both direct and indirect implications on human health. These effects are anticipated to increase current health risks, particularly in the most vulnerable communities. Currently, many tiny island states in the Indian Ocean experience climate-related health issues, such as extreme weather-related morbidity and mortality, as well as some vector-borne and water-borne diseases. Tropical cyclones, storm surges, flooding, and droughts are examples of extreme weather conditions and climate events that can have both immediate and long-term implications on human health. The majority of SIDS in the Indian Ocean are located in tropical regions with adverse effects of climate which can lead to diseases like malaria, dengue, and others. The IPCC assessments have repeatedly demonstrated that a lack of access to sufficient, safe freshwater and enough nourishment can substantially jeopardise human health on islands. Additionally, there is rising concern in island communities in the Caribbean Sea, Pacific, and Indian Oceans that a lack of freshwater and more severe storms and droughts could result in a decline in standards of hygienic practices (McMichael & Lindgren, 2011).

The most likely global climate change would jeopardise human security on a number of fronts is now widely acknowledged. Extreme weather events would cause disruptions in people's lives, food supplies would become scarcer, and coastal communities would move inland as a result of the IOR. The region's fisheries industry, which is vital to the blue economy, has been harmed by marine pollution, illicit fishing, and overfishing, which have led to an already diminishing resource base. It is a concern for achieving sustainable blue economy agendas that as the effects of climate change on seas and marine life continue to become more apparent, the problems already present in the industry are likely to get worse.

Oceans could become warmed and deoxygenated as a result of climate change, which could have a negative impact on the coral reef ecosystems that serve as many fish species' habitats. Coral reefs have already suffered terribly as a result of rising ocean temperatures. For instance, the coral reefs in the Indian Ocean were primarily affected by the global mass bleaching of 1998, which led to a 46 per cent average reduction in coral cover in the area. An increase in ocean surface temperature also makes it possible for zooplankton and phytoplankton composition to shift, which lowers fish production.

Sea-level rise, extreme precipitation events, tropical cyclones, and storm surges have already had an impact on small island villages and coastal cities. In addition to affecting economics, culture, and water and food security, climate change also has an impact on human settlements and infrastructure, health, and wellness (high confidence). Given that a large portion of their population, infrastructure, and economic resources are concentrated in the low-elevation coastal zone below 10 metres elevation, these changes are a big worry for tiny islands. Atoll islands exhibit the challenges of growing exposure and vulnerability the most.

Even with a low greenhouse gas emission scenario (high confidence), the vulnerability of communities in tiny islands, particularly those who depend on coral reef systems for a living, may surpass adaptive limitations well before 2100. The vulnerability of low-lying and coastal areas to the effects of climate change poses major concerns regarding the ability of the land to sustain human life and means of subsistence. Although the reasons and results of climate-related migration are very context-specific, and there is not enough data to estimate the number of climate-related migrants today and in the future, it is predicted to rise.

Small islands have already begun to suffer losses and damages, especially as a result of tropical cyclones and sea-level rise. The methodologies and mechanisms to quantify climate-induced losses and damages for small islands remain substantially underdeveloped despite the loss of human life and economic damage. Furthermore, there are just a few reliable approaches for attribution assessments.

Adaptation and Mitigation Strategies in Indian Ocean Islands

In the coastal regions of the Indian Ocean basin, climate change will exacerbate environmental and socio-economic issues. Climate change also poses a serious threat to human health, water resources, natural ecosystems, forestry, agriculture, and food security. The priority sectors for East Africa are coastal areas, agriculture, forestry, and water resources.

The most significant issue for coastal nations across the region is the rise in sea levels. The repercussions of climate change would fall disproportionately on the Least Developed Countries (LDCs), especially the Small Island Developing States (SIDS). They make up a small portion of the global greenhouse gas emissions, but they also suffer additional natural risks and have a limited ability to ameliorate the effects of global warming. Maldives, only 3 metres above sea level, faces the terrifying possibility of being fully flooded by fast-rising oceans within the next century. The Intergovernmental Panel on Climate Change (IPCC) has identified small island states, low-lying coastal cities, and mega deltas like the Ganges-Brahmaputra and Mekong as climate change 'hotspots' in the Indian Ocean Region, making them particularly susceptible to the effects of rising sea levels (IPCC, 2014).

The emphasis of this section is on adaptation for coastal zones, even though both mitigation and adaptation are crucial strategies for addressing the climate change challenge. A broad range of meanings and implications exist for the concept of adaptation. The IPCC uses a broad definition of adaptation that includes changes in natural or human systems to present or anticipated climatic stimuli or their effects that mitigate harm or take advantage of advantageous opportunities. Adaptation can be proactive (taking action before effects are noticed) or reactive (responding to the effects of climate change). The ability to adapt to climate change is referred to as adaptive capacity, which also provides an indication of the various capacities of societies to respond to climate pressures on a technological, financial, institutional, etc., basis. Developing nations and LDCs require outside assistance because they have lesser adaptive capacities than developed nations. Adaptation policy refers to increasing adaptive capacity and developing basic infrastructure to meet the challenges.

Conclusion and the Way Ahead

Several enablers can be used on the Indian Ocean Islands despite the presence of adaptation barriers to enhance adaptation outcomes and increase resilience. Governance and legal reforms, public policy measures, and various types of cultural knowledge resources are some of these enablers.

The stakeholders in the Indian Ocean Region face numerous adaptation policy issues. A major barrier to effective adaptation measures in Asia, according to a United Nations Framework Convention on Climate Change (UNFCCC) study, is a dearth of trustworthy data. Discrepancies and discontinuities in recorded data, as well as a wide range of data-gathering techniques, processing, and reporting for effects and vulnerability assessments, frequently thwart the establishment of integrated policies. Time-series data is frequently insufficient, and monitoring equipment failure is another major worry (Plaza, 2007). As a result, systematic monitoring for a variety of oceanographic and atmospheric variables that are required to analyse changing trends and evaluate new policy needs is lacking for many coastal locations.

Another significant barrier is a lack of money. According to the UNFCCC, international financing is required to promote adaptation in developing nations as part of the 'polluter pays principle', albeit specific funding levels have not yet been established. The Global Environment Facility, LDC Fund, Special Climate Change Fund (SCCF), and the Adaptation Fund are just a few of the funding options, but current investment is still in the early stages.

There is a lack of availability of data on adaptation methods and the scope. Some of the SIDS countries have gradually developed a knowledge base to make themselves resilient in the wake of climate-related challenges. It is necessary to have a better knowledge of the restrictions on adaptation as well as what, in small island environments, defines current resilience and/or effective adaptation. Greater understanding of the factors that reduce local and indigenous resilience, as well as awareness of the socio-political environments in which communities operate and the decision-making processes, can help identify opportunities at all scales to improve climate adaptation and facilitate action toward pathways for climate-resilient development.

Climate change is a very important issue to address in contemporary times. Water bodies, particularly islands, are very much critical with the climate change. The Indian Ocean island countries are impacted both socially and economy wise. The causes of climate change have diverse effects in terms of sea levels. The ocean ecosystem suffers significantly in different Indian Ocean island countries. To protect against further damage due to climate change, mitigation and adaptation strategies are needed.

References

Dasgupta, S. (2007). *The impact of sea level rise on developing countries: A comparative analysis* (Vol. 4136). World Bank Publications.

Haider, R., Rahman, A. A., & Huq, S. (Eds.). (1991). *Cyclone'91: An environmental and perceptional study*. Bangladesh Centre for Advanced Studies.

IPCC. (2014). *Summary for policymakers. Climate change 2014: Impacts, adaptation, and vulnerability. Contribution of Working Group II to the fifth assessment report of the Intergovernmental Panel on Climate Change*. Field, C. B., V. R. Barros, D.

J. Dokken, K. J. Mach, M. D. Mastrandrea, T. E. Bilir, M. Chatterjee, K. L. Ebi, Y. O. Estrada, R. C. Genova, B. Girma, E. S. Kissel, A. N. Levy, S. MacCracken, P. R. Mastrandrea, and L. L. White (eds.), (pp. 1–32). Cambridge University Press, Cambridge, United Kingdom and New York, NY.

McMichael, A. J., & Lindgren, E. (2011). Climate change: Present and future risks to health, and necessary responses. *Journal of Internal Medicine*, *270*(5), 401–413.

Nicholls, R. J., Wong, P. P., Burkett, V. R., Codignotto, J. O., Hay, J. E., McLean, R. F., Ragoonaden, S., & Woodroffe, C. D. (2007). Coastal systems and low-lying areas. In M. L. Parry, O. F. Canziani, J. P. Palutikof, P. J. van der Linden, & C. E. Hanson (Eds.), *Climate change 2007: Impacts, adaptation and vulnerability. Contribution of Working Group II to the fourth assessment report of the Intergovernmental Panel on Climate Change* (pp. 315–356). Cambridge University Press.

Plaza, H. C. (2007). *Background paper impacts, vulnerability and adaptation to climate change in Asia*. Available: https://unfccc.int/sites/default/files/alam_back_paper.pdf

Webb, A. P., & Kench, P. S. (2010). The dynamic response of reef islands to sea-level rise: Evidence from multi-decadal analysis of island change in the Central Pacific. *Global and Planetary Change*, 72(3), 234–246.

8 Climate Refugees in the Indo-Pacific

Developmental and Geopolitical Aspects

Faisal Ahmed and Pushkar Pushp

Introduction

The impact of climate change on displaced people is enormous. According to the United Nations High Commissioner for Refugees (UNHCR), an average of 21.5 million individuals were forcefully displaced by sudden weather-related hazards from 2008 to 2016, while thousands more were displaced by gradual climate change-related hazards.[1] The actions taken by policymakers will affect the number of individuals who are displaced due to climate change events. Nevertheless, tens of millions of individuals are expected to be displaced due to climate change-related impacts in the next 20–30 years.[2] In fact, it is estimated that there will be around 216 million internal climate migrants by 2050 (Huang et al., 2022).

Asia and the Pacific are the most vulnerable regions to be impacted by climate change. Bangladesh, the Maldives, and Tuvalu are facing threat of climate change (ADB, 2012). For instance, in Bangladesh, increased tidal levels, cyclones, and continuous erosion have made the lives of the coastal people vulnerable. Due to unabated erosion, approximately 22 per cent of people from Sandwip Island were forced to migrate to the city of Chittagong and other parts of the country. In fact, due to climate-related issues, residents of Kutubdia Island had to leave their homes. The study reveals that to protect themselves from erosion, floods, and other climate catastrophes, around 50,000 climate-displaced people were living in Cox's Bazar city (Barua et al., 2017; Ciavola et al., 2015). In fact, in the Pacific Islands where the rising sea level is submerging several islands and causing people to move to bigger countries, it is estimated that by 2100, almost 50 islands will be lost to the ocean. Moreover, in South Asia, climate change is deteriorating the existing migration patterns due to rising temperatures, sea levels, cyclones, flooding, and extreme weather events. Economic growth and urbanisation are facing the repercussions of climate change, particularly in low-lying coastal areas that are already vulnerable due to the rise in sea level (ADB, 2012). A prediction by the World Bank estimated that approximately 800 million people would face severe repercussions due to climate change, and the economies of South Asia might lose 1.8 per cent of their annual GDP due to the ongoing catastrophe of climate change by 2050.

DOI: 10.4324/9781003586869-9

A study by ADB (2012) also outlines that to tackle the ongoing issues with the people who are displaced due to climate change, participants of the Paris Agreement requested the Warsaw International Mechanism for Loss and Damage Associated to develop recommendations to deal with the people who are affected. The impact of climate change worsens the living conditions of those who are displaced and also affects their living conditions.[3] Moreover, basic amenities such as water are in scarce supply to serve the host refugees.

Rainfall, droughts, and desertification are rising along with the rise in sea levels. These events have forced people to cross borders and move to other countries. Considering the challenges these people face, refugee and human rights law should ascertain that climate-displaced people are given protection. People who are displaced by climate change often find themselves in areas called 'hotspots' where it becomes increasingly difficult for them to return home. The population residing in coastal regions vulnerable to the impact of rising sea levels has risen from 160 million to 260 million over the last 30 years, with 90 per cent of these people coming from poorer, developing nations and small island states. Climate refugees are often not classified as 'refugees', as per UNHCR. This classification does not warrant them legal protections, and they are unable to defend their human rights against threats like deportation. For instance, the asylum application of Ioane Teitiota, a resident of Kiribati, was rejected by New Zealand. He had applied for 'refugee status' in New Zealand due to climate-related challenge in Kribati. Moreover, he also approached the UN Covenant on Civil Liberties; however, the committee defended the decision taken by New Zealand government.

Climate refugees are not protected under the 1951 Convention Relating to the Status of Refugees. Given the growing problem of climate refugees, it is essential to establish a clear definition of climate refugees and gather comprehensive data on Internally Displaced Persons (IDPs). An international mechanism should be created to protect these individuals.

This chapter thus assesses the factors that are causing the climate refugee crisis in the Indo-Pacific region, and how UNHCR and other multilateral bodies are concerned about the ongoing catastrophe.

Conceptual Background

The term 'climate refugee' as well as 'climate migrants', 'environmental migrants', 'climate-induced displacement', 'climigration', 'eco-migrants', etc. have all been used in the literature to refer to persons who have been uprooted because of the effects of climate change (Hartmann, 2010; Naser, 2011; Staszkiewicz, 2014; Naser et al., 2019). A standardised definition of climate-induced migration does not exist, though. The term 'climate refugee' is not recommended by UNHCR, and more accurately it refers to 'persons displaced in the context of disasters and climate change'.[4]

Environmental refugees, according to Myers (1997), are those who feel extremely unsafe in their own countries due to climate risks such as droughts, floods, and deforestation and have little choice but to migrate to a safe location with little to no hope of returning (see also Berchin et al., 2017). Ahmed and Mishra (2020) outline three reasons, viz. rising sea levels, extreme weather events, and water scarcity as the primary reasons for climate-induced migration. Environmental refugees are 'those people who have been forced to leave their traditional habitat, temporarily or permanently, because of a marked environmental disruption (natural and/or triggered by people) that jeopardize their existence and/or seriously affects the quality of their life' (ElHinnawi, 1985, pp. 4–5). Berchin et al. (2017) cited El-Hinnawi (1985) and Jacobson (1988) and they argued that there exist three different categories of environmental refugees. The first set of people faces repercussions on a temporary basis, i.e., they are temporarily displaced and they return after the area under disaster gets renewed. The second set of people migrates permanently due to climate disturbances caused by human actions. Finally, the last category of people migrates due to climate change for better living (Berchin et al., 2017).

The Internal Displacement Monitoring Centre (IDMC) reveals that around 255.6 million people were displaced due to climatic disasters worldwide between 2008 and 2019 (IDMC, 2020). [5] The estimates revealed that the number of people being displaced each year ranges from 10 million to 26 million. Climate change is believed to be a bigger reason for displacement than war and persecution combined (Perez, 2006; Faber & Schlegel, 2017). The World Health Organization estimates that 225 million people are impacted by natural disasters. These statistics demonstrate the gravity of the challenge climate migration poses and an urgent need for attention (Laczko and Aghazarm 2009). In fact, the projections related to climate refugees are more disturbing. For instance, IPCC predictions estimate the displacement of up to 1 billion people by 2050 (Warner et al., 2008).

According to the Norwegian Refugee Council, climate-related abrupt natural disasters uprooted as many as 20 million people in just 2008 alone (Elverland, 2009; Islam & Shamsuddoha, 2017). Between 2008 and 2013, approximately 165 million individuals experienced displacement due to sudden-onset disasters such as earthquakes, floods, landslides, and tropical storms. Additionally, in 2015, a Category 5 tropical cyclone, unprecedented in its impact on the Pacific Islands of Vanuatu, affected 166,000 people, leaving 75,000 without adequate shelter. The nexus of climate-related hazards, compounded by factors such as inadequate human development, accelerated urbanisation, and poverty, synergistically contributes to the phenomenon of forced migration (Islam & Shamsuddoha, 2017).

Faber and Schlegel (2017) outline that climate change negatively affects the life of poor people by eroding their limited resources, i.e., physical capital, which includes damage to shelter and infrastructure; human capital which includes a high rate of disease and malnutrition; social capital damage in the form of disruption of beneficial social networks; natural capital which includes damage to agriculture and marine ecology; and financial capital, which included damage to income and property. Marginalised people often lack sufficient resources to recover from the shock of climate change (see also Smith, 2006). For instance, Cyclone Aila in Bangladesh in 2009 destroyed 234,000 homes and displaced 123,000 people. The Gorkha earthquake in Nepal in 2015 displaced 2.6 million people. The drought in Afghanistan in 2018–19 affected two-thirds of its population and displaced 260,000 people. Moreover, these climate-induced disasters damaged agricultural fields, houses, and schools.

Cyclones, floods, and droughts often lead to the loss of agricultural land, loss of wages, and destruction of property. Often, climate disasters take a toll on people's health, wealth, and education. In search of better opportunities, they migrate. However, migration does not always lead to a better life for many. Due to their already poor economic background and limited skills, they are left isolated. For instance, a study by Ahsan (2019) outlines that due to their lack of experience outside of farming and fishing, climate change migrants were forced to engage in a variety of economic activities to survive in the urban environment. In fact, the climate migrants often find places near the urban periphery for their survival. These slums are not well managed and lack proper drainage and sanitation systems. The migrants are poor, and it becomes difficult for them to settle in a proper urban setting.

Moreover, the 1951 Geneva Convention does not provide grounds for refugee status to climate-induced migrants (Apap, 2019). This makes the climate refugees more vulnerable to migration. The current global policies for migration do not adequately address the issue of climate-induced migration (Wilkinson et al., 2016). As a result, managing climate migration is becoming a policy concern at both national and international levels. Kawajiri (2018) highlights the importance of developing regional and international frameworks to address climate-induced migration. For instance, the African Union's 'Kampala Convention', which was introduced in 2009 as the first legal project to assist internally displaced people in Africa. Another initiative, the 'Nansen Initiative', proposed by Norway and Switzerland in 2012, seeks to tackle potential legal and protective deficiencies concerning individuals compelled to migrate as a consequence of climate change.[6]

In 2019, around 24 million people were displaced due to climate change. Estimates reveal that people from 140 countries were displaced, with the Pacific region being affected significantly.

Small island countries like Fiji, Kiribati, Tonga, Tuvalu, and Vanuatu are vulnerable due to the effects of climate change (Burkett, 2011; IDMC, 2020; Ahmed & Mishra, 2020). Perumal (2018) reveals that Vanuatu is more exposed than any other country in the world (Comes et al., 2016). Cyclones, flood, storm, landslides, and other disasters like tsunamis make this region vulnerable to climate change. As a result of climate change, there is a disruption in crop production and food security. Moreover, due to the rise in the sea level, the low-lying coastal areas are susceptible to coastal erosion (see also Comes et al., 2016). Rising sea levels and storm surges create a catastrophe for other low-lying lands in the Pacific Islands, including Kiribati and Tuvalu. In fact, the study by Merone and Tait (2018) predicted that both the Pacific Islands Kiribati and Tuvalu will become uninhabitable by 2050.

All Pacific Islands (except Papua New Guinea) are surrounded by the sea, and a majority of the population lives in the vicinity of the coastal areas. Apart from land security, coastal communities are also vulnerable to livelihood security due to reduced crop production and coral reef degradation associated with ocean acidification (ILO, 2014).[7] Prolonged periods of droughts are also a likely source of climate change migrants, especially the highland region of Papua New Guinea, which suffers from drought and severe frost (ILO, 2014).

As per the Pacific Climate Change Science Programme (PCCSP), the Pacific region will witness an increase in the average temperature and extreme rainfall events. Moreover, issues pertaining to ocean acidification and sea-level rise will also be present. Though droughts are estimated to occur rarely, their severity will be very high. Further, tropical storms and cyclones are estimated to decrease, but their intensity will increase (ILO, 2014).

Not only in the Pacific countries but in Asia also, climate-induced migration is on the rise. For instance, in the context of Bangladesh, a study by Barua et al. (2017) reveals that between 1980 and 2016 around 370,000 people from three regions, namely Kutubdia, Maheskali, and Sandwip, migrated due to cyclones, floods, and coastal erosion. In the case of disaster, people's homes in the areas nearby their homes are destroyed. This further leads to permanent displacement from their native origin. This is quite common on islands, and the migrants have to search for everything: shelter to live, food to eat, and income to survive. Since most of the migrants are poor, it again becomes difficult for them to get employment in an urban environment. The poor mostly rely on natural resources for their livelihoods, which are vulnerable to environmental degradation due to climate change. For instance, a rise in salinity or persistent flooding depletes the fertility of the soil, destroying crops and endangering both biodiversity and livelihoods (Akter, 2009).

Climate-Related Challenges

Rise in Sea Level

Pacific Island will witness a rise in sea level ranging from 25 cm to 58 cm by the middle of this century, which will negatively affect the islands. A rise in two degrees Celsius of global temperature will also make marine life vulnerable. Moreover, the study by Ahmed and Mishra (2020) outlines that a rise in sea level can lead to coastal inundation and an increase in the number of natural disasters as well. A study by UNEHS (2017) reveals that in Kiribati itself, almost every household (94 per cent) was affected by natural calamities in the last 10 years, and around 81 per cent of the households were impacted by the rise in sea level. The main migration in Kiribati is from the outer islands to South Tarawa, the capital, which already is facing existential issues of overcrowding and water scarcity.

Moreover, in Tuvalu, 9 per cent of migration from 2005 to 2015 was caused by environmental conditions. Interestingly, the majority of households in the study expressed that 'they would consider migration as a necessary strategy if their basic living conditions were further worsened by the impacts of climate change'. The study noted that the most probable environmental factors that would trigger future migration are sea-level rise followed by saltwater intrusion, drought, and floods. Biermann and Boas (2010) projected that, by the 2080s, approximately 82 million individuals in Africa, 141 million in Asia, and 2.5 million in small island states would face the consequences of storm surges associated with an anticipated global sea-level rise of 38 cm (see also, Mallick & Vogt, 2014). It is also estimated that a rise in sea levels will threaten around 57 per cent of the infrastructure of the Pacific Islands in this century.[8] Several countries are facing the challenge of forced displacement due to floods and coastal erosion. The governments of Vietnam, Vanuatu, and Papua New Guinea have initiated relocation strategies for the exposed population.

Natural Disaster

Of the 281 natural disasters that took place worldwide in 2018, almost half of them occurred in the Asia-Pacific region. For instance, in Indonesia, two tsunamis and an earthquake occurred in quick succession, which affected the lives of many people. Shockingly, Japan also witnessed, unprecedented flooding and a heatwave that killed over 300 people. Moreover, South Asia experienced an unusual cyclone; Cyclone Ockhi, for instance, created catastrophe in the region. In Southwest Asia, a storm corridor of sand and dust collided with heavy thunderstorms and rain, causing widespread devastation across several countries. In the context of Bangladesh, a study by Mallick and Vogt (2013) reveals that Khulna and Satkhira districts were most affected by Cyclone Aila in 2009. In the four upazilas of these two districts, namely Dacope, Koyra, Shyamnagar, and Assasuni, around 234,000 houses were destroyed, and people were forced to migrate. Due to the scarcity of food and water, people started migrating to nearby cities. The study further outlines that around 123,000 people were displaced, the majority of whom belonged to Dacope and Koyra. Not only in Asia, but Pacific Islands too have witnessed several natural calamities in the past which have forced the population to migrate to other locations. For instance, Super Typhoon Yutu in 2018 destroyed several homes in Saipan and Tinian.

Over the past six decades, more than 2,400 tropical cyclones have occurred in the Pacific region, with an average of about 41 per year. Both the North and South regions of the Pacific experience frequent tropical cyclones. While the North Pacific feels the impact of tropical cyclones throughout the year, the South Pacific is affected between October and May. Tropical storms have impacted numerous Pacific Island countries, leading to widespread destruction, severe casualties, and high loss to economy.[9]

Coastal Erosion

Approximately 2.4 billion individuals in Asia inhabit low-lying coastal zones and flood plains. Seventeen out of the 25 cities vulnerable to a one-metre sea-level rise are situated in the Asia-Pacific region. In Bangladesh, 130 million people reside at an elevation of approximately 1.5 metres above sea level, experiencing frequent floods and storms exacerbated by inadequate drainage systems and infrastructure (Palinkas and Palinkas 2020). In the Pacific Islands, there is an anticipation of increased migration from low-lying islands to higher elevation regions like Fiji and Vanuatu (Palinkas and Palinkas 2020). The Government of Kiribati has implemented measures, including 'pre-emptive voluntary migration pathways', particularly through labour migration, and the acquisition of 20 square kilometres of land in Fiji for potential future refugee. [10,11]

Climate change affects land degradation and livelihood in some of the Pacific regions. For instance, low-lying coastal areas might become vulnerable to physical settlement. Moreover, reduced precipitation can cause the deterioration of habitability. Climate change poses a threat to the territorial integrity of certain Pacific Islands, including issues such as coastal inundation, river plain flooding, or erosion. In these instances, the physical locations designated for settlement may become unsuitable and incapable of supporting human habitation (ILO, 2014).

Coastal erosion is a widespread issue, as it affects coastlines across the globe and has a significant impact on both natural and human communities. The loss of coastal land to seawater has led to the displacement of people and the destruction of marine ecosystems. For instance, in India, 235 square kilometres of land was destroyed between 1990 and 2016 due to coastal erosion, leading to the displacement of millions of people. Though policies pertaining to rapid-onset disasters like monsoons and cyclones are present, disasters like coastal erosion are yet to be addressed at the policy level.[12]

Climate Change and Conflict

A report by UNHCR (2020) outlines that 95 per cent of all conflict displacements in 2020 occurred in countries which were vulnerable to climate change. Marginalised communities who have been displaced due to conflict caused by the effects of climate change are at significant risk, particularly in regions with inadequate governance and dispute resolution mechanisms. Climate change also worsens resource scarcity, causing conflicts and inducing migration of vulnerable populations who seek livelihoods. Moreover, changes in biodiversity also threaten food and economic security. The movement of large numbers of people due to climate-related challenges can bring different cultural groups together, leading to shifts in power balance, resource scarcity, and escalation among previously separated groups. When climate-related migrations occur near populated areas, or in locations that are crucial for political or economic activities, such as coastal zones, the threatening effects of climate change may have massive impacts overall.

Discussion and Conclusion

Regardless of the circumstances, people displaced by climate change may experience major human rights abuses both before and after relocation (Levy & Patz, 2015; Naser et al., 2019). A study by Naser et al. (2019) mentions that prior to displacement, natural disasters pose a threat to life, liberty, and security. When disasters with unexpected and rapid onset, like floods and cyclones, occur, the impacted population suffers harm and loss of life, homes and infrastructure are damaged, and crops are washed away. In addition, the inability of displaced people to return home or their long-term homelessness poses a major risk of human rights violations. It becomes a challenge for displaced people to get refugee status in other countries. For instance, the case of Ioane Teitiota, whose refugee status was rejected by the Government of New Zealand. He had applied for 'refugee status' in New Zealand due to climate-related challenge in Kribati. Moreover, he also approached UN Covenant on Civil Liberties; the committee defended the decision taken by New Zealand government.[13] New Zealand later introduced an 'experimental humanitarian visa' for climate-displaced people from Pacific Island countries. However, the initiative was dropped, and the plan for issuing 'climate refugee' visas was cancelled.[14] As outlined earlier, climate refugees are not covered by the 1951 Geneva Convention. Moreover, 'climate refugee' as a term is not recommended by the UNHCR; more precisely it refers to 'persons displaced in the context of disasters and climate change'.[15]

Policymakers must understand that climate change is a long-term issue with immediate implications on conflict, governance, and human rights.[16] The UNHCR is providing support to individuals who have been forcibly displaced within their own countries and are unable to return safely to their homes. The organisation has the capability to deploy teams to aid these displaced individuals through activities such as registration, documentation, and the provision of essential services, including shelter, basic hygiene, and nutrition. A report by the Asia-Pacific Foundation of Canada highlights four challenges pertaining to the preparedness to assist climate refugees.[17]

Inadequate international legal framework: The current international legal framework does not adequately address climate-induced migration, as the 1951 Geneva Convention and the 1992 UN Framework Convention on Climate Change do not directly incorporate this issue.

Inadequate disaster preparedness and risk management: There is a lack of long-term solutions, with many programmes providing only short-term relief to climate-displaced communities rather than long-term solution. There is also a lack of resources for improvement in basic infrastructure such as healthcare, and infrastructure to reduce migration issues.

Limited insurance penetration: Financial strategy for disaster relief is inadequate, especially in developing countries. In fact, the report highlights that while 40 per cent of loss is covered by disaster risk insurance in developing countries, it is only about 5 per cent in Asian countries. Moreover, insurance policies hardly cover climate-induced migration.

Urbanisation: Rapid urbanisation in Asia, which is largely driven by migration, may intensify the challenges posed to climate refugees, particularly in densely populated cities that lack the necessary infrastructure to accommodate migrant inflows. Bangladesh has implemented a migration strategy that encourages rural-urban migration, but finding sustainable solutions to increased climate-induced migration in cities remains a major challenge.

A report by UNHCR on the Strategic Framework for Climate Action outlines three pillars for action to assist the affected communities. The pillars are:

1 *Law and Policy*: It may include actions such as supporting the states for the development and implementation of a legal and institutional framework to facilitate displacement due to disaster. Moreover, the policy supports international protection for those in need, which also includes refugees.
2 *Operations*: This objective is in line with the SDG 2030 agenda principle, which is 'leaving no-one behind'. The outcome of this objective is to provide energy needs (cooking, lighting, and heating) to refugee households in emergencies. Moreover, it aims to provide sufficient access to sustainable energy for cooking along with basic lights and connectivity.
3 *UNHCR environmental footprints*: This objective is in line with reducing the harm to the environment by the efficient use of natural and financial resources.[18]

A report by McAllister (2023) suggests that creating economic opportunities in societies which are vulnerable to environmental change can help these communities. For instance, in Bangladesh, due to floods salinity of 53 per cent of farmland has increased, becoming a threat to the community that relies on agriculture.[19] With support from the Dutch research project Salt Solution and local NGOs, farmers have been able to grow salt-tolerant crops, including potatoes, carrots, cabbages, and coriander. Approximately 10,000 farmers were assisted, resulting in extra harvests per year.[20]

In fact, organisations such as the World Bank are also helping the countries with risk reduction in climate-affected areas. The World Bank is helping Samoa make its main transportation infrastructure more resilient. Additionally, along with the UN, it is assisting coastal communities to become more resilient. In fact, the World Bank has also provided support to Samoa for recovery from Cyclone Evan, including repairing damaged roads and bridges and providing resources such as farming equipment and seeds to the affected farmers. In Vanuatu, the World Bank along with the EU is promoting climate-resilient crops and livestock among farmers, as well as implementing disaster risk management programmes. Alongside Japan, the World Bank is installing a tsunami warning system in Port Vila and Luganville and is also enhancing Vanuatu's national hazard response systems.[21]

The importance of Multilateral Development Banks (MDBs) has been mentioned in the study of Huang et al. (2022). The study highlights that projects related to climate adaptation, along with migration and climate-induced displacement, should be discretely tailored based on local climate and migration conditions. The study mentions that MDBs invest in sustainable projects, especially in low- and middle-income countries. In 2021, MDBs committed USD 81 billion to climate finance, and some of these projects also includes the projects on climate-induced migration and displacement. In fact, MDBs are significantly putting their efforts into tackling climate-induced migration in two ways, first, by making large-scale investments in the infrastructure in locations that are either experiencing or might experience climate displacement. For instance, the World Bank supports urban flood resilience in 11 projected hotspots of climate migration in Niger. Second, by funding the technical and financial tools. For instance, the World Bank and ADB facilitated an early warning system for weather predictions.

Climate-induced migration needs a cooperative approach from the respective agencies and countries. Partnership is necessary to support climate-vulnerable countries (Huang et al., 2022). Moreover, it is not only the responsibility of agencies or MDBs to facilitate investment; rather, the state should also incorporate financing and policy guidelines for climate refugees. Nevertheless, legal acceptance of climate refugees remains vital, especially for those who cannot return to their homes due to climate disasters (Berchin et al., 2017).

Notes

1 Report on the Impact of Climate Change on Migration (October 2021), https://reliefweb.int/report/world/report-impact-climate-change-migration-october-2021
2 https://www.zurich.com/en/media/magazine/2022/there-could-be-1-2-billion-climate-refugees-by-2050-here-s-what-you-need-to-know
3 https://www.unhcr.org/uk/what-we-do/how-we-work/environment-disasters-and-climate-change/climate-change-and-disaster
4 https://www.unhcr.org/uk/what-we-do/how-we-work/environment-disasters-and-climate-change/climate-change-and-disaster
5 https://www.nzaia.org.nz/ghoshandorchiston.html#:~:text=In%20the%20context%20of%20Pacific,2011%3B%20IDMC%2C%202020b).
6 https://www.nzaia.org.nz/ghoshandorchiston.html#:~:text=In%20the%20context%20of%20Pacific,2011%3B%20IDMC%2C%202020b).
7 https://www.ilo.org/dyn/migpractice/docs/261/Pacific.pdf
8 https://edition.cnn.com/2021/07/08/us/pacific-islands-climate-change-refugees/index.html
9 https://www.worldbank.org/content/dam/Worldbank/document/EAP/Pacific%20Islands/climate-change-pacific.pdf
10 https://www.unescap.org/sites/default/files/GCMPREP_5E.PDF

11 Ibid.
12 https://reliefweb.int/report/india/climate-change-displacement-and-managed-retreat-coastal-india
13 https://racetozero.unfccc.int/climate-refugees-the-worlds-forgotten-victims/
14 https://www.nzaia.org.nz/ghoshandorchiston.html#:~:text=In%20the%20context%20of%20Pacific,2011%3B%20IDMC%2C%202020b).
15 https://www.unhcr.org/uk/what-we-do/how-we-work/environment-disasters-and-climate-change/climate-change-and-disaster
16 https://www.usip.org/publications/2021/09/how-climate-change-deepens-bangladeshs-fragility
17 https://www.asiapacific.ca/sites/default/files/climate_refugees_v4.pdf
18 https://www.unhcr.org/uk/what-we-do/how-we-work/environment-disasters-and-climate-change/climate-change-and-disaster
19 https://www.zurich.com/en/media/magazine/2022/there-could-be-1-2-billion-climate-refugees-by-2050-here-s-what-you-need-to-know
20 Ibid.
21 https://www.worldbank.org/content/dam/Worldbank/document/EAP/Pacific%20Islands/climate-change-pacific.pdf

References

ADB. (2012). *Addressing* climate change *and* migration *in Asia and the Pacific.* Available at: https://www.adb.org/sites/default/files/publication/29662/addressing-climate-change-migration.pdf

Ahmed, F., & Mishra, V. (2020). Estimating relative immediacy of water-related challenges in Small Island Developing States (SIDS) of the Pacific Ocean using AHP modeling. *Modeling Earth Systems and Environment*, *6*(1), 201–214.

Ahsan, R. (2019). Climate-induced migration: Impacts on social structures and justice in Bangladesh. *South Asia Research*, *39*(2), 184–201.

Akter, T. (2009). Climate change and flow of environmental displacement in Bangladesh. Centre for Research and Action on Development, Dhaka.

Apap, J. (2019). *The concept of "climate refugee": Towards a possible definition.* https://www.europarl.europa.eu/RegData/etudes/BRIE/2018/621893/EPRS_BRI(2018)621893_EN.pd f

Barua, P., Rahman, S. H., & Molla, M. H. (2017). Sustainable adaptation for resolving climate displacement issues of south eastern islands in Bangladesh. *International Journal of Climate Change Strategies and Management*, *9*(6), 790–810.

Berchin, I. I., Valduga, I. B., Garcia, J., & de Andrade, J. B. S. O. (2017). Climate change and forced migrations: An effort towards recognizing climate refugees. *Geoforum*, *84*, 147–150.

Biermann, F., & Boas, I. (2010). Preparing for a warmer world: Towards a global governance system to protect climate refugees. *Global Environmental Politics*, *10*(1), 60–88.

Burkett, M. (2011). In search of refuge: Pacific islands, climate-induced migration, and the legal frontier. *Asia – Pacific Issues, 98*, 1–8.

Ciavola, P., Uddin, M. M., Duo, E., Lee, B., & Fakhruddin, S. H. M. (2015, December). Vulnerability of Bangladesh coastline to inundation under cyclone activity: Past records and DRR strategies at Sandwip Island. In E-proceedings of the 36th IAHR World Congress, The Hague (pp. 45-46).

Comes, M., Dubbert, M., Garschagen, M., Hagenlocher, M., Sabelfeld., Yew Jin Lee, U., Grunewald, L., Lanzendörfer, M., Mucke, P., Neuschäfer, O., Pott, S., Post, J. Schramm, S.,Schumann-Bölsche, D., Vandemeulebroecke, B., Well, T., &

Birkmann, J. (2016). World Risk Report. Berlin & Bonn: Bündnis Entwicklung Hilft / ited Nations University - EHS.

El-Hinnawi,E.(1985).*Environmentalrefugees*.UnitedNationsEnvironmentProgramme. http://rfmsot.apps01.yorku.ca/glossary-of-terms/ environmental-refugee/

Elverland, S. (2009, June). 20 million Climate Displaced in 2008, Norwegian Refugee Council. available at: http://www.nrc.no/?did=9407544.

Faber, D., & Schlegel, C. (2017). Give me shelter from the storm: Framing the climate refugee crisis in the context of neoliberal capitalism. *Capitalism Nature Socialism*, *28*(3), 1–17.

Hartmann, B. (2010). Rethinking climate refugees and climate conflict: Rhetoric, reality and the politics of policy discourse. *Journal of International Development: The Journal of the Development Studies Association*, 22(2), 233–246.

IDMC. (2020). *GRID: 2020 global report on internal displacement*. https://doi.org /10.4324/9780203965436

ILO (2014). The employemt impact of climate change adaptation. Available at: https://www.ilo.org/sites/default/files/wcmsp5/groups/public/@ed_emp/documents /publication/wcms_645572.pdf

Islam, M. R., & Shamsuddoha, M. (2017). Socioeconomic consequences of climate induced human displacement and migration in Bangladesh. *International Sociology*, *32*(3), 277–298.

Islam, M. M., & Shamsuddoha, M. D. (2018). Coastal and marine conservation strategy for Bangladesh in the context of achieving blue growth and sustainable development goals (SDGs). *Environmental Science & Policy*, *87*, 45–54.

Laczko, F., & Aghazarm, C. (2009). Migration, Environment and Climate Change: Assessing the Evidence. Geneva: International Organization for Migration. https:// environmentalmigration.iom.int/migration-environment-and-climate-change -assessing-evidence

Jacobson, J. (1988). Environmental refugees: A yardstick of habitability. *Bulletin of Science, Technology & Society*, *8*, 257–258. http://dx.doi.org/10.1177 /027046768800800304

Kawajiri, K. (2018). Protection of cross-border climate displaced persons in the South Pacific: Case of Tuvalu and New Zealand. *International Public Policy Research*, 22(2), 21–43.

Lawrence, H., Sohst, R., & Le Coz, C. (2022). *Financing responses to climate migration*. Available at https://www.migrationpolicy.org/sites/default/files/ publications/mpi_climate-dev-banks_final.pdf

Levy, B. S., & Patz, J. A. (2015). Climate change, human rights, and social justice. *Annals of Global Health*, *81*(3), 310–322.

Mallick, B., & Vogt, J. (2014). Population displacement after cyclone and its consequences: Empirical evidence from coastal Bangladesh. *Natural Hazards*, *73*(2), 191–212.

McAllister. (2023). https://www.zurich.com/en/media/magazine/2022/there-could-be -1-2-billion-climate-refugees-by-2050-here-s-what-you-need-to-know

Merone, L., & Tait, P. (2018). Preventing disaster in the Pacific Islands: the battle against climate disruption. *Australian and New Zealand Journal of Public Health*, *42*(5), 419–420.

Myers, N. (1997). Environmental refugees. *Population and Environment*, *19*, 167–182. http://dx.doi. org/10.1023/A:1024623431924

Naser, M. M. (2011). Climate change, environmental degradation, and migration: A complex nexus. Wm. & Mary Envtl. L. & Pol'y Rev., *36*, 713.

Naser, M. M., Swapan, M. S. H., Ahsan, R., Afroz, T., & Ahmed, S. (2019). Climate change, migration and human rights in Bangladesh: Perspectives on governance. *Asia Pacific Viewpoint*, *60*(2), 175–190.

Palinkas, L. A., & Palinkas, L. A. (2020). Fleeing flooding: Asia and the Pacific. In *Global climate change, population displacement, and public health: The next wave of migration* (pp. 99–125). Springer, Cham. https://doi.org/10.1007/978-3-030-41890-8_6

Perez, Teresita. 2006. "Climate Refugees: The Human Toll of Global Warming." Center for American Progress. http://www.americanprogress.org/issues/2006/12/climate_refugees.html

Perumal, N. (2018). "The place where I live is where I belong": community perspectives on climate change and climate-related migration in the Pacific island nation of Vanuatu. *Island Studies Journal*, *13*(1), 45–64.

Smith, D. M. (2006). *Just one planet: Poverty, justice and climate change*. Soluciones Practicas.

Staszkiewicz, Z. (2014). The Impact of Contemporary Environmental Migrations on International Security. *Przegląd Strategiczny*, (7), 343–357.

UNEHS. (2017). Kiribati: Climate Change and Migration Relationships Between Household Vulnerability, Human Mobility and Climate Change. Available at: https://collections.unu.edu/eserv/UNU:5903/Online_No_20_Kiribati_Report_161207.pdf

UNHCR. (2020). https://www.unhcr.org/neu/wp-content/uploads/sites/15/2021/08/10-facts-about-refugees-2021-ENG_Vi2.pdf

Warner, K. et al. (2008). "Field observations and empirical research. Climate change and displacement". *Forced Migration Review*, *31*, 13–14.

Wilkinson, E., Kirbyshire, A., Mayhew, L., Batra, P., & Milan, A. (2016). *Climate-induced migration and displacement: Closing the policy gap (Issue October)*. https://www.odi.org/publications/10594-climateinduced-migration-and-displacement-closing-policy-gap

9 Climate-Induced Forced Migration in South Asia

Rajendra Prasad Patel and Anju

Introduction

Environmental changes in general and those related to climate change in particular are now widely acknowledged as major global drivers of migration. As a result of the unavoidability of these effects, mitigation efforts alone will not be sufficient to combat climate change; they must be accompanied by adaptation strategies. By making people and communities more resilient to these effects, adaptation aims to lessen the effects of climate change. Despite the fact that they were originally thought of as two potential choices, mitigation and adaptation measures are increasingly understood as being necessary to combat climate change (ADB, 2012, p. 9).

Although environmental disasters that cause human displacement have drawn more attention recently, it is still unclear how communities will respond in reality to long-term environmental change. Many people believe that there is a causal link between climate change and migratory patterns, and that all populations in areas where there is climate change would be compelled to relocate. Long-term ecological change has resulted in a developing migration driver. The impact of climate change on human displacement will be highlighted by environmental change. Climate-related migration streams will be entangled with larger-scale migration components; as a result, they shouldn't be seen as being constrained. This work places a high priority on understanding ecological relocation as a component of global change, which is also a prerequisite for effective migration and adaptation policy (ADB, 2012, p. 12).

In South Asia, large populations are often uprooted as a result of floods, dry spells, soil degradation, tropical storms, and typhoons. A disproportionate number of deaths, relocations, and injuries associated with such events affect the poor. They are least prepared to adapt when their houses and communities are damaged by extreme weather because they are forced to live in low-lying shoreline deltas, riverbanks, floodplains, steep slants, and degraded urban environments where the effects are typically severe. The region is anticipated to experience further significant social, political, and financial changes in the coming decades, in addition to being significantly impacted by environmental change. These numerous changes will likely

DOI: 10.4324/9781003586869-10

have an effect on migratory patterns, from environmental change to more affordable travel. The type and volume of population mobility will also be influenced by public strategies, such as adaptation processes and migration management (World Bank, 2021).

More than half of all South Asian countries, encompassing around 750 million people in eight countries – Afghanistan, Bangladesh, Bhutan, India, Maldives, Nepal, Pakistan, and Sri Lanka have been impacted by at least one environment-related calamity over the most recent 20 years. South Asia is one of the most climate-affected regions in the world. The South Asian peoples are living through a 'new climate normal', where intensifying heat waves, cyclones, droughts, and floods are challenging the constraints of governments and peoples to adapt, and these difficulties are simply likely to deteriorate (World Bank, 2021). All countries in the region face more serious dangers from environmental change and are less prepared to embrace them. The McKinsey Institute in its study said that 'More than 62 billion people in South Asia, including 45 million in India alone will be forced to migrate from their homes due to climate disasters by 2050' (The Times of India,2020).

Theoretical Understanding

Essam Hinnawi coined the term 'Environmental refugees' in his report prepared for the United Nations Environment Programme in 1985. He defined them as 'those people have been forced to leave their traditional habitat, temporarily or permanently, because of marked environment disruption, natural and/or triggered by people that jeopardise the existence and seriously affect the quality life'. According to this definition, one might distinguish three classifications of refugees, specifically: (i) those displaced temporarily due to an environmental pressure like an earthquake or a cyclone, who will return to their territory when conditions normalise; (ii) people who have permanently migrated due to a permanent change caused to their habitat, for example, due to the establishment and related man-made lakes; and (iii) the people who are displaced forever or temporarily in search of an improved quality of life because their original habitat cannot meet their basic needs (Ghosh, 2016, pp. 45–46).

The terms 'environmental refugee' or 'climate change refugee' appear often in the media and writing, which refer to people who are forced to leave their homes and cross national borders because of the increased occurrence of severe weather events or increases in extensive risk due to environmental issues. However, the term has no basis in global regulation. Under the Refugee Convention, the term 'refugee' is reserved for people who have been compelled to cross borders due to the threat of persecution; this definition is generally accepted to include people escaping from armed conflict (Holzer, 2012). The experts in this field have increasingly recognised and corrected the misuse of this term and now refrains from its utilisation (Renaud et al.,2011). Moreover, experts warn against the augmentation extension of refugee regulations to incorporate environmentally displaced people, as this would not 'solve' the humanitarian issue (McAdam, 2011).

The majority of climate-induced migrants and displaced people relocate to cities, though the proportion of rural versus urban destinations varies depending on the circumstances. Climate change and pressure, in the case of both extensive and intensive risk, typically affect entire communities rather than specific individuals or families. People who have been displaced in large numbers have been incorrectly referred to as 'climate change refugees' in the context of climate change (Wilkinson et al., 2016, p. 2).

Climate Change and Human Mobility

People move for a variety of reasons, and human mobility is divided into many modes and categories. Understanding the characteristics of these movements and their relationships to various climate shocks and stressors is critical for developing effective policy responses, adaptation plans, and investments. There is a fundamental gap in understanding the links between climate change and human mobility, particularly migration, and more research on this relationship is required to better understand the decision-making process. There is no universally accepted definition of 'climate-induced human mobility' (Warner, 2010), but it broadly refers to the movement of people caused by abrupt or gradual changes in weather or climate. This can include temporary and permanent movement, seasonal and singular movement, as well as voluntary and forced movement.

Displacement

Displacement refers to situations in which people are compelled to leave their homes or their regular residences. Displacement is typically linked to high risk when the main factor causing displacement is the occurrence of a disastrous event. It can occur within or outside of national borders (ADB, 2012, p. 9).

Migration

Movements that are somewhat voluntary carry a significant risk and can occur within or across national borders. The decision to relocate is complicated and frequently influenced by several factors, including but not limited to climate risk (ADB, 2012).

Forced Migration

Migration involves elements of coercion, such as risks to one's life and livelihood, whether resulting from natural or artificial causes (for example, movements of refugees and internally displaced persons, as well as people displaced by natural or environmental disasters, chemical or nuclear disasters, famine, or development projects). Although causes for moving may be mixed and there may be some human agency or volition involved in the decision to relocate, forced migration often suggests a lack of choice (World Bank, 2018).

Push and Pull factors

In the study of migration, push factors are those that motivate people to migrate away from their current location, and pull factors are those that entice people to relocate. For instance, Cyclone Amphan hit Bangladesh and India in May 2020. Three million people were forced to evacuate, and about two million homes were destroyed or damaged. It was one of the biggest storms to hit the area in decades. In Bangladesh, India, and Sri Lanka, three different nations, people were uprooted and lost their houses. Many of the destroyed homes were mud huts and decaying shacks, making them completely unprepared for disasters (Kugeman, 2020).

Cyclone Amphan is merely the most recent example of how climate change consequences and migration brought on by climate change are real and present problems in South Asia. The regional governments have developed strategies to lessen these dangers. Such attempts are hampered by capacity, governance, and funding limitations. They urgently require foreign assistance, which has so far been there but insufficient (Kugeman 2020).

When considering South Asia's vulnerability to climate change, many observers automatically focus on Bangladesh, a low-lying, lower riparian country that frequently experiences severe floods. In truth, the area as a whole is highly vulnerable. Flooding and rising sea levels pose a threat to Bangladesh's coastal regions as well as those of India, Pakistan, and Sri Lanka. The consequences of climate change significantly threaten the inhabitants of these countries, exacerbated by the size and density of their metropolitan coastal populations. Afghanistan, Bhutan, and Nepal, which are all landlocked, are currently dealing with rising temperatures, drought, and glacial melt. The Maldives, the world's lowest-lying nation and smallest but most densely populated island, faces the very real possibility of total submersion in the not-too-distant future.

It is not unexpected that at least one climate-related tragedy has had an effect on about 700 million people during the past 10 years. India and Pakistan are among the ten most susceptible nations to climate change, according to a Germanwatch think group. In addition, alarming research from India's Ministry of Earth Sciences, published in June 2020 and based on extensive climate modelling, forecasts that South Asia's most populous nation, India, will become significantly drier and hotter in the coming decades, with average temperatures on track to rise by nearly four degrees Celsius by the end of the century. Along with rising temperatures in the Indian Ocean and expected sea level rises of up to nearly a foot, it will also suffer longer monsoon periods, increased glacial melt, and hotter temperatures elsewhere (Kugeman, 2020).

Ineffective public policy exacerbates climate vulnerability on the Indian Subcontinent. Water shortages are made worse by government subsidies for wasteful flood irrigation and water-intensive crops like sugar. Developers and business owners are given unrestricted access to valuable water resources. In India, towns in at-risk areas have officials who ignore flood dangers except during election seasons.

Climate Change Migration in South Asia

South Asian nations are among the most susceptible to the consequences of climate change, which can range from short-term disruptions brought on by escalating extreme weather conditions like rain and heat to longer-term changes that can reach critical points like sea-level rise. Moving temporarily or permanently, domestically or abroad, is one adaptation technique. Short-term climate events may cause short-term migration, whereas longer-term migration for climate adaptation may be a response to longer-term changes (Conigliani et al., 2022). Due to local environmental changes, as many as 40.5 million people in South Asia could migrate internally by 2050, with 19.9 million of them living in Bangladesh (Clement et al., 2021).

As climate change affects the environment and labour demand in destination nations, cross-border migration patterns may likewise shift. Remittances from cross-border migration enable households to insure against environmental shocks at home. For Nepal, Pakistan, and Sri Lanka, whose remittances are above the South Asian average of 4.2 per cent of GDP, changing cross-border movement patterns could be difficult (World Development Indicators, 2020). For equitable, sustainable development, policymakers must understand how migration is impacted by climate change and how that mobility affects productivity. One of the main climate risks to the area is water. The majority of flood-exposed individuals reside in South and East Asia, and a substantial percentage of the poor population lives in both South Asia and Sub-Saharan Africa (Rentschler et al.,2022). People have adapted in areas at danger of seasonal flooding by diversifying their lifestyles to prevent migration (Chen & Mueller, 2018) or by relocating temporarily (Mobarak & Reimo, 2020). Even when they have evolved, harsh conditions may still be beyond their capacity, which can alter seasonal migration patterns. Seasonal migration to rural areas may be disrupted because potential migrants might prefer to stay with their family rather than leave, for example, in the event of severe flooding (Mobarak & Reimo, 2020). The limited mobility of these impoverished households may expose them to additional risks.

Similar to flooding, droughts reduce agricultural productivity and cause more people to leave rural regions (Dallmann & Millock, 2017; Zaveri et al., 2021). The poorest people deal with water shortages, fewer economic opportunities in their local region, and a lack of means to locate opportunities elsewhere because they are 80 per cent less likely to migrate (Zaveri et al., 2021). Water and migration are related, which emphasises the need to address the distributional effects of water risk and related policies, such as water management in agriculture.

Additionally, South Asia has experienced rising average temperatures, which raises the risk of intense heat. These occurrences boost rural migration and lower agricultural yield (Mani et al., 2018). More migration is linked to rising temperatures than to flooding and other natural calamities (Baez et al., 2017; Mueller et al.,2014). Similar to how heat makes air pollution worse, recent research indicates that pollution is linked to skilled worker exodus and decreased overall output (Khanna et al., 2021).

One of the long-term effects of climate change that the area will experience is sea-level rise. This is a gradual process, and households adapt without using migration first (Hauer et al., 2020). Greater internal and cross-border migrations, as well as increased aquaculture, are linked to the gradual rise in soil salinity in coastal Bangladesh than actual flooding is (Chen & Mueller, 2018, 2019). Because families are affected differently and have various resources to migrate, it will be essential to understand the thresholds that drive migration as a result of increasing sea levels when developing policies to address the potential distributional impact.

The anticipated increase in zoonotic illnesses represents another risk associated with climate change. The majority of climate migrants will relocate to urban areas, where the risk of infectious disease spread is higher due to inadequate sanitation and a lack of public amenities. Disease transmission is therefore likely to worsen (Lall et al., 2021). The COVID-19 epidemic emphasises how crucial it is to increase urban resilience to address climate change (World Bank, 2021). The pandemic also emphasises how migration plays a part in disease spread (Lee et al., 2021a). A risk exposure indicator has been developed in Bangladesh using bilateral migratory connections with COVID-affected areas (Ahsan et al., 2020).

The unexpected implications of lockdowns, the most often used policy option, are also demonstrated by recent studies. Lockdowns forced South Asian migrant workers to return home, and evidence from Bangladesh, India, and Pakistan suggests that while international movement is linked to community spread sown by migrants (and others) returning from outside, internal migration is a weaker predictor of infection (Lee et al., 2021a). In Bangladesh and Nepal, the epidemic and the government's response to it also caused earnings to decline and food insecurity to rise more among migrant households than non-migrant households (Barker et al., 2020).

Due to the lack of prospects at home and the fact that most migrants work in low-paying professions abroad, these negative consequences may continue to affect migrant households. A long-term plan, such as compiling a database of migrant workers and creating a more skilled labour force, should be used in conjunction with the short-term strategy of providing a social safety net for migrant workers (Karim et al., 2020).

Understanding these migrants' characteristics and the effects they have on production at both the destination and the place of origin is crucial because climate change continues to have an impact on migration. The trade-off between encouraging migration and spending on programmes to support locals in areas with significant climate risks should also be taken into account by policymakers. Social protection can lessen the need to migrate, but limiting migration as a method for adaptation could harm potential migrants' welfare and raise the cost of catastrophe preparedness (Dasgupta et al., 2022). To safeguard livelihoods in urban and rural areas, policies that prioritise no-regret solutions like increasing urban resilience (Lall et al., 2021), providing market access for coastal households , agricultural insurance (Mobarak & Reimo, 2020) should be prioritised.

Impact of Climate Change in the Region

In the Sunderbans, a UNESCO World Heritage Site that is home to the biggest mangrove forest in the world, the high risk of climate-related migration in South Asia is particularly evident. It is extremely vulnerable to sea-level rise, catastrophic storms, land erosion, and high salinity of the ocean due to its location along the Bay of Bengal and its proximity to Bangladesh and India. Storms have caused human emigration from the Sunderbans Islands in recent years. Given that important employment sectors like farming, fishing, betel leaf cultivation, and tourism have been negatively impacted by weather-related occurrences, more flight may be on the horizon.

In the worst-case scenario—one in which the region lacks climate-friendly policies—a 2018 World Bank study forecasts that by 2050, there will be close to 40 million climate migrants in South Asia. According to the World Bank's forecast, in this case, close to a quarter of all internal migrants in South Asia and up to 2 per cent of the region's population would be categorised as climate migrants. By 2050, the Bank predicts that there would be up to 20 million climate migrants worldwide, even in the best-case scenario with numerous climate-friendly initiatives. The Delhi-Lahore corridor, which connects India and Pakistan, as well as coastal cities in both India and Bangladesh, are among the out-migration 'hotspots' predicted by the World Bank. The territories of Nepal and the mountains of southern India will see the most in-migration activity.

Bangladesh is anticipated to be the starting point for regional climate migration. In the worst-case scenario, more climate migrants than any other category of internal migrants might be living in the nation by 2050—more than 13 million in total.

Due to the catastrophic floods, the 9.4 million acres of crops that have been impacted.). The following planting season was impacted, which may have an influence on future food supply chains, due to the uncertainty around when the water would recede. Due to food shortages and rising vegetable prices, inflation is already at risk of increasing (Mangi & Dilawar, 2022).The damage caused by Pakistan's recent floods has once again highlighted South Asia's vulnerability to climate change and the urgent need to increase climate resilience. In the region, environmental degradation and natural disasters will worsen due to climate change (UNEP, 2016). In the coming decades, the effects of climate change will combine with other stressors, including pollution and resource overuse, to have an impact on a population that is both urbanising and expanding quickly (World Bank, 2018).

In other words, some will experience forced moves, displacement, or the need for relocation due to the effects of climate change. People who are relocating are drawn to hospitable situations (Hunter et al.,2015). On the other hand, people are as likely to move to environments that are stressful for the environment (Foresight, 2011). Millions of individuals will be 'stuck' in environmentally stressful locations because they are unable or unwilling to leave those areas, rendering them immobile or 'trapped' (Foresight 2011).

Conclusion

In the face of climate hazards, people and communities will turn to various migration-related adaptation approaches, such as permanent migration, temporary migration, or in situ adaptation, depending on the extent, measure, or level of susceptibility. Migration may not be permanent due to social variables and perceptions of risk, even in the case of slow but irreversible changes in the environment, such as rises in temperature or sea level. Regarding future climate change, particularly the extent and magnitude of impacts and social responses to reduce and adapt to changes, there is still a significant deal of uncertainty. Impacts will be a complex mix of little annoyances and serious disruptions that, in some situations, will result in poverty and fatalities, necessitating sophisticated remedies.

There will be many different ways that migratory patterns may change, but they almost certainly involve an increase in the push forces (those at the origin) that cause emigration pressures. Three primary mechanisms are likely to lead to their manifestation:

1. a rise in the likelihood of environmental risks and the socio-ecological events they cause;
2. shifting resource circumstances in linear and nonlinear trends that affect how easily natural resources may be accessed and used; and
3. climate change affects perceptions, independent of actual experiences.

Millions of people are already displaced each year in South Asia due to extreme natural disasters. Numerous of these move away. Their numbers will increase, and their movement patterns will shift over time, especially as human settlements are affected by slow-onset environmental changes like drought and sea-level rise. Sadly, international collaboration to address the cross-border aspects of such migration is still insufficient for the task at hand and the issues that have yet to materialise. Therefore, it is critical that all governmental levels take proactive measures to solve this issue through legislation, projects, and funding. Humanitarian disasters that may have been avoided will occur if the issues concerned are not given substantial, timely attention.

The displacement of entire populations will be less prevalent than migration as a means of adaptation to the effects of climate change. When all other options for adaptation and community resilience have been exhausted and moving becomes a question of survival, displacement will happen as a last resort.

The most vulnerable groups should have access to migration options, such as the choice to stay put (and adapt) or move to a location that is safer and more secure. Greater public funding will be needed for responses in the areas of urban investments, catastrophe risk reduction, social protection, and livelihood development. Additionally, the private sector may contribute significantly to risk reduction. In large measure, the policies and investments decided upon today will determine whether or not migration brought on by climate change manifests as forced migration and displacement or as a planned and aided adaptation strategy. All levels of government, as well as regional and international organizations, have a responsibility to act with urgency on this matter. They have an opportunity to shape a future of human mobility that encourages better, more inclusive living conditions for the inhabitants of the South Asian Region.

The effects of climate change have shown that many nations lack policies and processes for implementing systemic shocks to migration. In order to assist displaced migrants, governments are forced to quickly develop and put into action assistance and reintegration measures. This knowledge could be used to create more resilient migration policies that would not only assist migrants and their families in surviving such shocks in the future but also encourage migration by lowering the risks associated with migrating. One of the most important components of de-risking migration policies is close cooperation between receiving and sending regions.

If coordinated climate and development actions are not implemented, the size of internal climate migration will increase by 2050 and then accelerate. Countries might anticipate 'hotspots' of migratory in- and outflow caused by the climate. This will have a significant impact on how nations plan their future development. If properly managed and supported by sound development policies, focused investments, and careful management, migration can be a rational climate change adaptation option. Internal climate migration may be a fact, but a catastrophe is not necessarily required. Reducing the number of persons compelled to migrate in misery could be accomplished by taking action in three key areas. First, reduce greenhouse gas emissions immediately; second, integrate climate change into development planning; and thirdly spend money now to better understand internal climate migration.

References

ADB. (2012). *Addressing climate change and migration in Asia and Pacific.*

Ahsan, R., Iqbal, K., Khan, M., Mobarak, A. M., & Shonchoy, A. (2020). *Using migration patterns to predict COVID-19 risk exposure in developing countries* [Policy brief]. Yale School of Management.

Baez, J., et al. (2017). Heat exposure and youth migration in central America and the Caribbean. *American Economic Review, 107*(5), 446–450.

Barker, N. C., et al. (2020). *Migration and the labour market impacts of COVID-19* (World Institute for Development Research Working Paper 2020/139). United Nations University.

Chen, J., & Mueller, V. (2018). Coastal climate change, soil salinity and human migration in Bangladesh. *Nature Climate Change, 8*(11), 981–985.

Chen, J., & Mueller, V. (2019). Climate-induced cross-border migration and change in demographic structure. *Population and Environment, 41*(2), 98–125.

Clement, V., et al. (2021). *Groundswell part 2: Acting on internal climate migration.* World Bank.

Conigliani, C., Costantini, V., & Finardi, G. (2022). Climate-related natural disasters and forced migration: a spatial regression analysis. *Spatial Economic Analysis, 17*(3), 416–439.

Dallmann, I., & Millock, K. (2017). Climate variability and inter-state migration in India. *CESifo Economic Studies, 63*(4), 560–594.

Dasgupta, S., et al. (2022). Coastal dilemma: Climate change, public assistance and population displacement. *World Development, 150,* 105707.

Dilawar, I., & Mangi, F. (2022). Deadly Floods Inundate Farms in Pakistan, Flushing Away Crops. Available at: https://www.bloomberg.com/news/articles/2022-08-31/deadly-floods-inundate-farms-in-pakistan-flushing-away-crops

Foresight. (2011). *Migration and .global environmental change: Final project report.* https://www.gov.uk/government/uploads/system/uploads/attachment_data/ file/287717/11-1116-migration-and-global-environmental-change

Ghosh, P. S. (2016). *Migrants, refugees and the stateless in South Asia.* Sage Publication.

Hauer, M. E., et al. (2020). Sea-level rise and human migration. *Nature Reviews Earth & Environment, 1*(1), 28–39.

Holzer, V. (2012). *The 1951 refugee convention and the protection of people fleeing armed conflict and other situations of violence.* UNHCR Division of International Protection Legal and Protection Policy Research Series.

Hunter, L. M., Luna, J. K., & Norton, R. M. (2015). Environmental dimensions of migration. *Annual Review of Sociology, 41,* 377–397.

Karim, M. R., et al. (2020). COVID-19's impacts on migrant workers from Bangladesh: In search of policy intervention. *World Development, 136,* 105123.

Khanna, G., et al. (2021). *The productivity consequences of pollution-induced migration in China* (NBER Working Paper 28401). National Bureau of Economic Research.

Kugelman, M. (2020). *Climate-induced displacement: South Asia's clear and present danger.* Wilson Center.

Lall, S., et al. (2021). *Pancakes to pyramids: City form to promote sustainable growth.* World Bank.

Lee, J. N., et al. (2021). Migration, externalities, and the diffusion of COVID-19 in South Asia. *Journal of Public Economics, 193,* 104312.

Mani, M., et al. (2018). *South Asia's hotspots: The impact of temperature and precipitation changes on living standards.* World Bank.

McAdam, J. (2011). *Climate change displacement and international law.* The Office of the United Nations High Commissioner for Refugees.

Mobarak, A. M., & Reimão, M. E. (2020). Seasonal poverty and seasonal migration in Asia. *Asian Development Review, 37*(1), 1–42.

Mueller, V., et al. (2014). Heat stress increases long-term human migration in rural Pakistan. *Nature Climate Change, 4*(3), 182–185.
Renaud, F. G., et al. (2011). A decision framework for environmentally induced migration. *International Migration, 49*(1), 5–29.
Rentschler, J., et al. (2022). Flood exposure and poverty in 188 countries. *Nature Communications, 13*(1), 1–11.
The Times of India. (2020, December 18). Climate-induced migration set to treble in South Asia by 2050, India to face displacement of 45 million people. https://timesofindia.indiatimes.com/india/climate-induced-migration-set-to-treble-in-south-asia-by-2050-india-to-face-displacement-of-45-million people/articleshow/79802062.cms
UNEP (2016). The rise of environmental crime. Nairobi: UNEP.
Warner, K. (2011). *Climate change induced displacement: Adaptation policy in the context of the UNFCCC climate negotiations.* United Nations High Commissioner for Refugees (UNHCR), Division of International Protection.
Wilkinson, E., Leigh Mayhew, Pandora Batra and Andrea Milan (2016). *Climate-induced migration and displacement: Closing the policy gap.* ODI Briefing.
World Bank. (2018). *Groundswell: Preparing for international climate migration.*
World Bank. (2021). *Climate change action plan 2021–2025: South Asia roadmap.*
World Bank. (2022). *South Asia economic focus (Fall 2022) coping with shocks migration and the road to resilience.*
World Development Indicators. (2020). *Personal remittances, received (% of GDP) – South Asia.* https://data.worldbank.org/indicator/BX.TRF.PWKR.DT.GD.ZS?locations=8S
Zaveri, E., et al. (2021). *Ebb and flow, volume 1: Water, Migration, and Development.* World Bank.

10 Climate Change Vulnerabilities

Issues and Policy Responses in Bangladesh

Aparajita

Introduction

Bangladesh is a South Asian country that is extremely vulnerable to climate change due to its geographical location, nature of rivers, monsoons, increased precipitation during monsoon, high-intensity cyclones, rise in the sea level, etc. (Asaduzzaman et al., 2005). Bangladesh is among the ten most affected countries vulnerable to climate change in the world (Climate Risk Index, 2021). The social, economic, and physical characteristics of Bangladesh make it one of the most vulnerable countries to climate change (McCarthy et al., 2001). During the period from 2000 to 2019, it witnessed extreme weather conditions in about 185 events and climate change attributed cost accounted for $2.86 trillion (CRI, 2021) (WEF, 2023). Bangladesh consists of several waterways, including Ganga River, Brahmaputra, and the Meghna (Huq & Ayers, 2008). The waterways connect the country with the rest of the world, especially South Asia, provide fertile agricultural soil, and enhance travel in this area. Most of the country is floodplain; in the northwestern region, the sea level rising above 30 metres of mean level makes Bangladesh vulnerable to flooding, especially in the Northwest, Northeast, South Central, and Central regions of Bangladesh.

The geographical location of Bangladesh makes it vulnerable to floods, cyclones, sea-level rise, and other natural disasters (Huq, 2001). It can be seen that 30 to 40 per cent of Bangladesh region is flooded each year (Ahmed & Haque, 2002). It has been reported that 56 per cent of Bangladesh's total population live in 'high climate exposure areas', a major proportion are subject to 'very high' exposure (Ahmed & Haque, 2002). The location of Bangladesh is such that it is less than 15 feet above sea level, leading to rising sea levels, which threatens the lives of the people (MOEF, 2009). Bangladesh occupies a total area of 147,570 square kms (BBS, 2020). It is part of the Bengal Basin, which is covered with sediments and washed down mostly by the Himalayas. The problem increases when three major rivers[1] bring sediment loads to the region, leading to significant congestion. The prediction of the displacement of one in every seven people by 2050 due to sea-level rise would lead to the loss of land and migration in large numbers (Khatun & Sadat, 2021). 'Climate change in Bangladesh has started what may become

DOI: 10.4324/9781003586869-11

the largest mass migration in human history...Some scientists project a five-to-six foot (sea level) rise by 2100, which would displace perhaps 50 million people' (Glennon, 2017). The Sundarbans[2] face threats as coastal forests fail to sustain livelihoods and biodiversity.

Three main seasons can be witnessed in Bangladesh. (i) Extreme summer season that exceeds 40 degrees for almost ten days. The heavy rate of evaporation with heavy and irregular rainfall from March to June affects Bangladesh (Huq & Ayers, 2008). (ii)The heavy rainfall with humid and hot monsoon with temperatures between 20 and 36 degrees Centigrade (Huq & Ayers, 2008). (iii) The months of November to March are cooler and drier with temperature from 8 to 15 degrees Centigrade (Huq & Ayers, 2008). The temperature is generally high, rainfall is heavy, humidity is high, with different variations in seasons. .

The general surface warming of 0.3 degrees Centigrade to 5 degrees Centigrade will occur by the year 2050 and there will be increase in rainfall by about 5 per cent to 20 per cent (Mahtab, 1989). Around 44.36 per cent of all households were affected by natural disasters in 2015 (Khatun & Saadat, 2021).

Bangladesh was declared eligible to graduate from the category of Least Developed Countries to Developing Country in the year 2018. According to the United Nations Organisations, the criteria for graduation from one category to another depend on Gross National Income (GNI), Human Assets Index (HAI), and Economic Vulnerability Index (HVI) (UNO, 2022). For graduating to the category of Developing Country, the GNI should be 1230 USD or above, the HAI score should be 66 or above, and the Environmental Vulnerability Index (EVI) should be 32 or below. During the Financial Year 2016–17, the GNI per capita of Bangladesh was 1610 USD, the HAI score of Bangladesh was 72.9, and the EVI score of Bangladesh was 24.8. If the EVI score is focused on, it can be seen that Bangladesh has a seven-point margin of the EVI threshold. If the climate change issues in Bangladesh increase its environmental vulnerability, then there can be an increase in its EVI score, which can affect its gradation to the Developing Country category by 2024 (UNO, 2022).

Impacts of Climate Change in Bangladesh

Extreme Flooding

The increased monsoon rains with flash floods[3] together with a high rate of precipitation, result in rainwater floodings. The coastal floodings[4] are frequent due to the rise in the sea level. There have been strong downpours and a rise in temperature that melt Himalayan glaciers. The melts flow into rivers, making Bangladesh prone to devastating floods. According to the UNICEF (2021),

> Around 12 million of the children most affected [by climate change] live in and around the powerful river systems which flow through

> Bangladesh and regularly burst their banks. The most recent major flooding of the Brahmaputra River in 2017 inundated at least 480 community health clinics and damaged some 50,000 tube wells, essential for meeting communities' safe water needs.

The water flows from upstream catchment areas, together with heavy rainfall during monsoon, floodplain areas, a complex system of drainage, and storms in coastal areas, are the reasons for different types of floods in Bangladesh (Ahmed & Mirza, 2000). There are different types of natural floods, such as rainwater floods, river floods, coastal floods, and flash floods caused by storms (Ahmad et al., 2000). The coastal areas are negatively affected due to vulnerabilities caused by climate change, together with economic, social, and environmental issues. The involvement of stakeholders, spatial dynamics, and a mix of climate variables complicate the adaptation to climate change in coastal areas of Bangladesh (Rahman & Alam, 2003). The intensity and frequency of climate change in Bangladesh have increased over the years. Every decade, there has been increase of 0.20 degrees Celsius in the average temperature and a 153 mm increase in rainfall in Bangladesh (Rahman & Alam, 2003).

Issue of Groundwater and Farming in Bangladesh

The groundwater levels are largely affected due to climate change and subsequently lead to the utilisation of surface water, which is scarce and costly. As a result, small farmers have recourse to aquaculture. Agriculture dominates the coastal region, and 60 per cent of people are dependent on it for their living. One such example is the Cyclone Alia of 2009, when it was observed that most shrimp producers of shifted to crop production. However, the saline water has also affected crop production to a large extent resulting in its decline, and subsequently farmers look towards aquaculture.

Droughts and Sedimentation[5]

The lack of moisture in soil that is not sufficient to meet the needs of crops causes drought, and it is basically an agricultural phenomenon. It mainly occurs during the pre-monsoon period caused due to variability of hydrometeorology and the post-monsoon period due to lack of optimum rainfall (Karim et al.,1998). The rise in temperature together with low rainfall causes drought. The sediments are deposited by rivers mostly on flood plains.

Salinity Increase

The rising sea level also brings salinisation[6] that infiltrates agricultural land and hinders the growth of crops as their ability to take up water is affected. This salinisation also affects the drinking water of almost ten million people in communities residing in coastal areas, and consumption of such water affects the health of people, leading to cardiovascular diseases. In 1973, 8.3 million hectares were affected by saline water that increased to 105.6 million hectares in 2009 (Bangladesh's Soil Resources Development Institute). The coasts of Bangladesh are majorly affected by salinity, where the major portion of the country's population resides. Future predictions indicate that increases in floodings, droughts, and salinity would greatly affect agriculture, livestock production, aquaculture and fish production, forests, vegetation, and shrimp production. As climate change affects water-related disasters, the livelihood of the people would be greatly affected (RVCC, 2003).

Effects of Cyclones and Storms on Different Sectors

There are several sectors in Bangladesh that are most affected by climate change vulnerabilities. The industrial sector, fisheries sector, agriculture sector, navigation sector, domestic and municipal purposes, and flow of streams that control salinity are all impacted. It has been predicted that in the next twenty-five years, with the increase in population, the availability of water per person will substantially decrease (Ahmad et al., 2001). This will affect water availability during the dry season. The floods, droughts, erosion of river banks, sedimentation, and increase in salinity are caused by water-related problems. There are chances that poverty will increase by about 15 per cent due to climate change by 2030 due to the fall in crop productivity (Karim et al., 1996).

The area for rice production is greatly affected due to climate change vulnerabilities, especially cyclones and storms. The flood of 1988 diminished agricultural production by 45 per cent (Karim et al., 1996). The fish culture takes place in ponds and water tanks. The floods can negatively affect such fish culture and bring it under threat due to climate change vulnerabilities (GOB, 2005). The increase in salinity of water and high tides affects shrimp production (WB, 2000). The continuous floods affect the deaths of livestock (Ahmad et al., 2001; Choudhary et al. 2003) (Table 10.4). The season of droughts increases diseases among livestock. The effects on livestock would increase negatively even in future (Ahmed, 2005; GOB, 2005). The climate change would severely affect the Sundarbans forest, which is home to many species including the Royal Bengal tiger. The mangrove forest largely depends on the Ganga water supply. The inland penetration of saline water will affect the vegetation of forests. This will lead to the disappearance of high-quality scrubs and species.

Migration

The rise in sea level leads to floods, an increase in wetlands, shoreline erosion, and erosion at the river banks, especially in coastal areas. The livelihoods of people residing in these areas are greatly affected and ultimately lead to migration on a large scale. The continued intensity of disasters causes migration and refugee problems (Ahmed & Ahmed, 2002). The net migration in the year 2021 in Bangladesh was about minus 174,500 (2021) (World Bank, 2022). The male members of the family leave behind their female counterparts who lack social and economic security. Subsequently, frequent migration has been witnessed from rural areas to urban areas, especially Dhaka, Rajshahi, Khulna, and Chattogram. This has led to a rise in land surface temperature of about two degrees Celsius (Khatun & Saadat, 2021). The large-scale migration to urban areas affects the infrastructure and health of people in urban areas. The water level increases in the Ganges-Meghna-Brahmaputra River Basin affect livelihood to a large extent and has led to 10 million people becoming refugees. The rate of global warming depicts that 17 per cent of the people would have to be relocated (Khatun & Saadat, 2021). About 81 per cent of the people said that they migrated to the slums of Dhaka because of climate issues, and it has also been predicted that there would be 3 to 10 million migrations within Bangladesh in a span of 40 years as a consequence of climate change severity (GED, 2018).

Natural disasters have displaced around 700,000 Bangladeshis over the last decade every year (Internal Displacement Monitoring Centre, 2021). Bangladesh has faced powerful cyclones such as Cyclone Sidr (2007), Cyclone Aila (2009), Cyclone Roanu (2016), Cyclone Bulbul (2019), Cyclone Amphan (2020), and Cyclone Yaas (2021) (Ahmed 2006). The Cyclone Amphan was the strongest cyclone ever recorded in the history of Bangladesh, leaving many people homeless and destroying several districts and lands. The prediction is that such weather conditions would hamper and aggravate the lives of agrarian communities in Bangladesh. The Intergovernmental Panel on Climate Change has reported that the capacity to adapt, sense, and expose towards climate change determines the vulnerability level of an ecological and social system (IPCC, 2001).

Public Health

The health of the people has been negatively affected due to climate change. There is a lack of safe drinking water and malnutrition on a large scale among the affected population, especially migrants, both physically and mentally. There have been respiratory diseases, hypertension, and preterm births due to contamination of groundwater. The frequent cyclones and floods affect the infrastructure, healthcare services, food supplies, and the ability of the government to respond to such negative consequences of climate change (Table 10.1; Table 10.2). The use of nets around beds, and stockpiling of medicines and food are majorly practiced among the population to cope with the severe health problems caused by floods, cyclones, and climate change vulnerabilities (Table 10.2). People also maintain hygiene to prevent infections caused by food and water. Awareness programmes are also organised for people to be prepared for such situations. The activities of pathogens[7] are

Table 10.1 Climate change and people's health

Climatic variables	*Effects on people's health*
Extreme rise in temperature	Skin diseases, stroke, anxiety and mental disorders, dementia, hypothermia, headache
Irregular rainfall	Diarrhoea, cholera, cough, risk of snakes, skin problems
Saltwater intrusion	Skin problems, respiratory problems, miscarriage, diarrhoea, hypertension
Natural disasters	Cholera, diarrhoea, mental problems, malnutrition, infections, loss of lives
Waterlogging and floods	Cholera, diarrhoea, skin diseases, malnutrition
Variable precipitation	Respiratory and cardiovascular diseases
Vectors' excessive breeding	Visceral leishmaniasis
Overexploitation of ground water	Leukomelanosis, dorsum, gangrene, skin cancer, keratosis,
Stagnant water condition	Cardiovascular, respiratory, and allergy
Groundwater pollution	Nervousness, adverse effects on reproductive and endocrine systems
Shortage of food production	Malnutrition and deficiency

Source: Isfat and Raihan (2022)

Table 10.2 Health coping strategies

Issues	*Health coping strategies*
Access to healthcare	Deciding between qualified and unqualified providers
Mental health	Relatives, neighbours, and friends can help; self-medication and traditional practices must be avoided.
Extreme weather conditions	Advance preparation for food and medicines, post-disaster preparedness, preserving safe drinking water.
Vector-borne diseases	Avoiding being outdoors at the time of dawn and dusk
Food borne diseases	Hygiene maintenance during preparing and consuming food.
Water borne diseases	Prevention from algal bloom
Air quality disruption	Continuous warning and monitoring on high pollution days
Heatwaves/cold waves	Warnings and preparedness during emergencies

Source: Isfat and Raihan (2022)

strongly linked with climate variability (WB, 2000; Koelle et al., 2005). The events of El Nino[8] increase the cases of cholera (Rodo et al. 2002). Malaria and dengue cases also increase with the vulnerability of climate change. The National Adaptation Programme of Action (NAPA) considered that the effects of climate change on the health of people were not certain. The government expenditure on health has increased to minimise the adverse impact of climate change on people's health (Agrawal et al., 2003).

Infrastructure and Settlement

If the World Bank data on Bangladesh is analysed, it can be perceived that almost half of the population lives below poverty line (World Bank, 2022). The GDP of Bangladesh is 416.26 US$ billion (2021) (World Bank, 2022). The GDP per capita is around 2,457.9 US$ (2021) (World Bank, 2022). These data demonstrate that Bangladesh faces problems as an LDC, and it gets worse due to climate change impacts. The infrastructure, like improper and poorly built homes, is affected by natural disasters (Ahmed et al., 2004) (Table 10.3). The networks, roads, markets, educational institutions, and administrative buildings are devastated by floods with high density (Nizamuddin et al., 2001; Siddiqi, 1997; Siddique & Chowdhary, 2000). The cyclone of 1991 destroyed several embankments in coastal areas like Chittagong airport's runway, and several ships had drowned in Chittagong's Patenga Channel (Haider, 1992). The social and economic situations together with the well-being of the people are affected due to climate disasters. Many scholars are of the view that the country faces poverty due to issues caused by natural disasters (Mirza et al., 2003; Ahmad & Ahmad, 2002). These disasters need to be managed to achieve the goals of development in the country. The Government of Bangladesh has focused on the management of disasters to achieve developments (MOFDM, 2005). If the government does not take effective measures, climate change would cost around 121 billion USD during the period from 2005 to 2050 (Department of Economics, 2006). The social and economic situations, the geographical location, and exposure to climate change make the agrarian community most vulnerable in the world (Ramamasy & Bass, 2007).

Due to natural disasters in Bangladesh, the country has to go through economic loss, and this loss has increased over the years (Table 10.4). Bangladesh, as an LDC, has to face huge burden from such economic loss, placing the government under several debts to look after its population. In Table 10.3, the economic has been shown for the period from 2009–2014 and it has been predicted that the loss will increase in the coming years with the increase in frequency and intensity of natural disasters.

Gender

The women are found to be more vulnerable to climate change issues in Bangladesh. As women are mostly involved in household responsibilities, the shortage of water and food builds psychological pressure of feeding the family. They even have to cover long distances to fetch water for the family in case of water and food shortages. According to Alston and Akhter, until gender equality is achieved both at national and international levels, women will face such problems of food and water insecurity. The lack of proper sanitation and infrastructure facilities affects the women and teenage girls. They also face pregnancy issues like hypertension, eclampsia, and pre-eclampsia

Table 10.3 Household affected by natural disaster (2009–2014)

	Drought	*Flood*	*Water logging*	*Cyclone*	*Tornado*	*Storm/ tidal surge*	*Thunderstorm*	*River/ coastal erosion*	*Landslides*	*Salinity*	*Hailstorm*	*Other disasters*
National	14.80	34.48	13.88	21.31	04.14	08.65	14.94	04.95	00.08	04.09	11.88	07.90
Barisal	01.41	05.24	03.91	78.31	00.91	31.51	03.72	04.35	00.00	00.85	00.31	00.05
Chittagong	10.61	32.03	34.39	30.96	01.80	13.51	08.39	07.01	00.80	05.30	09.46	12.86
Dhakha	19.89	51.89	18.68	00.00	03.88	00.00	17.69	06.42	00.00	00.00	20.86	09.27
Khulna	09.30	07.68	34.88	23.23	02.62	09.16	07.39	04.15	00.00	22.24	10.31	07.32
Rajshahi	25.39	48.47	00.65	00.00	07.51	00.00	20.40	03.39	00.00	00.00	12.86	14.73
Rangpur	23.99	41.74	00.68	00.00	12.30	00.00	23.53	06.87	00.00	00.00	16.62	08.34
Sylhet	16.51	69.97	02.57	00.00	01.30	00.00	31.84	01.95	00.02	00.00	12.54	05.42

Source: Khatun and Saadat (2021)

Table 10.4 Economic loss due to natural disasters (2009–2014)

	All sectors	*Crops*	*Livestock*	*Poultry*	*Fishery*	*Land*	*House*	*Homestead & forestry*
All natural disasters	184247.34	66703.42	8772.16	2224.88	10713.93	49229.73	31676.89	14926.27
Drought	10569.20	9144.99	191.14	81.93	189.65	698.15	00.00	263.34
Flood	42807.19	22163.26	2373.29	593.81	1986.77	8966.45	5040.03	1683.58
Water logging	16062.24	8660.70	702.99	204.59	2466.43	1541.53	1769.64	716.36
Cyclone	28384.81	4194.25	3137.51	750.92	2109.46	00.00	10833.38	7359.29
Tornado	4299.03	984.46	145.80	28.02	00.00	00.00	1847.69	656.13
Storm on tidal surge	12676.02	2343.78	769.93	321.96	3271.24	3318.00	6212.35	803.42
Thunderstorm	10940.12	2493.63	432.28	103.39	00.00	00.00	2034.28	1698.47
River or coastal erosion	36408.92	1076.20	729.30	33.90	338.72	31742.12	21.44	454.40
Landslide	249.01	07.78	00.16	00.11	00.00	200.23	46.92	19.29
Salinity	6072.94	2162.73	142.23	11.01	00.00	2763.21		946.84
Hailstorm	11471.69	9679.63	53.38	27.01	00.00	00.00	1386.53	325.14
Other natural disasters	4306.11	3792.00	94.14	68.28	351.73	00.00	00.00	00.00

Source: Khatun and Saadat (2021)

especially in coastal areas. The climate change vulnerabilities and sickness along with lack of opportunity affect the ability of women to earn for the family.

Security and Climate Change

The security issue includes the aspects of 'freedom from want' and 'freedom from fear'. The 'freedom from want' aspect includes food, water, health, livelihood, and environment issues, and the 'freedom from fear' aspects include political, personal, and asset security. Therefore, the climate change vulnerabilities of Bangladesh question the 'freedom from want' aspect of security. (i) Food Security: People face loss of crops due to floods, drought, storms, salinity of water, and cyclones, and the threat is increasing every year. (ii) Water Security: The long-term drought and dry season affect drinking water. There has also been contamination of fresh water resources with the saline water, affecting drinking water. (iii) Life and Property Loss: There has been loss of assets, homes, cattle, and other livelihood essentials due to storms, floods, cyclones, and other natural disasters. (iv) Livelihoods Loss: Opportunities for people due to floods, cyclones, and storms are lost. Fishing opportunities are limited due to rough seas. Health problems and other issues limit people's daily opportunities. (v) Women Insecurity: Women are more vulnerable to such climate change vulnerabilities as they look after children, elderly people, and other members of their family. (vi) Displacement: People migrate to slums in urban areas, especially in Dhaka, and such influx has led to an increase in insecurity and crime in these areas.

Climate Change and Problems in Bangladesh

The question arises as to why the Bangladesh government has not been successful in dealing with the issues of climate change vulnerabilities. This section analyses the reasons behind unsuccessful attempts in controlling the adverse impacts of climate change. (i) There is a lack of political desire to deal with the issues, which prevents proper management of disasters caused by climate change. (ii) The ineffectiveness and failure of Pressure Groups[9] in Bangladesh is another important reason that prevents the effective functioning of the planning and policies concerning climate change issues. (iii) Bangladesh faces financial constraints[10] and resource constraints[11] to deal with weather issues as it is a Least Developed Country[12] and every year the country suffers economic loss to a large extent due to cyclones, floods, storms, and other climate-related vulnerabilities (Table 10.4). There are resource constraints in these countries in terms of social, economic, institutional, and technological aspects. Climate change affects migration, livelihood, and resources of poor people. (iv) There is a lack of knowledge and awareness among the people about the impacts of climate change. (v) The problems of weak and bad governance[13] enhance the devastating effects of climate change. (vi) The increased frequency of climate change does not give time and space for proper management of climate change. (vii) There has been a lack of cooperation among stakeholders[14] that creates problems in the management and execution of policies concerning climate change. (viii) There is a lack of institutional mechanism, proper planning, and implementation on the part of the government and other stakeholders concerning climate issues. (ix) Bangladesh lacks organisational capacity[15] that hinders the process of climate management, reconstruction, and rehabilitation. (x) Weak implementation and interlinkages of policies are other important issues that affect the working of the government. (xi) Specific sectors and issues are not addressed either by the government or Non-Governmental Organisation[16] or other institutions related to environment and climate change.

Policy Responses of Bangladesh both at National and International Levels

There is an absence of any comprehensive national policy from the government specifically focused on climate change issues. However, there are several policies and programmes adopted in Bangladesh that reduce the impacts of climate change, such as programmes related to the generation of employment, alleviation of poverty, diversification of crops, programmes that address issues concerning climate change, and events that deal with management and reduction of risks related to disasters. There are several programmes that discuss national policies of the Bangladesh government.

The Government of Bangladesh, through its Bangladesh Institute of Peace and Security Studies (BIPSS)[17] has established the Climate Change Cell under the Ministry of Environment and Forests. The Climate Change Cell regulates, monitors, and makes policies for the agencies of the government, policy formulators, and partners concerning foreign development (BIPSS, 2023). The Cell focusses on the security perspective of climate change. It studies different dimensions of climate change like human security, inter-state conflict, and social unrest. The Comprehensive Disaster Management Programme and the United Nations Development Programme have supported the Climate Change Cell in its work. There is assistance for creating a 'Climate Change Cell' within the Department of Environment to enhance the capacity building of the government concerning issues of climate change . The Cell works on raising awareness, advocates and suggests mechanism to climate change and for reducing risks in development activities.

There are several projects that address the shortage of food production, employment generation, development of community, improvement of infrastructure, and credit facilities. The rural poverty alleviation programme and employment generation programmes help in production and export of shrimps and fish. The programmes mentioned above have played a significant role in alleviating the disastrous effects of climate change (Kelkar & Bhadwal, 2007).

The Participatory Disaster Management Programme (PDMP) focusses on climate change adaptation, management, and prevention of disasters.

The emphasis is on preparedness that decreases the impacts of disasters. Measures like making people aware of different ways through which the negative effects of disasters are minimised. The knowledge, information, and skills in dealing with disasters at the national level need to be enhanced and an action plan in disaster-prone areas needs to be established. The measures for risk reduction of risks at local levels and early warning systems need to be improved (Kelkar & Bhadwal, 2007).

A Comprehensive Disaster Management Programme (CDMP) was established in 2003 in Bangladesh with the United Nations Development Programme (UNDP). The aim is to focus the attention of the government towards the reduction of risks and the preparation for disasters. There are several components of disaster management in the CDMP. It focusses on three major areas: (i) Building capacity for the Ministry of Environment and integrating climate change issues into their current working activities, (ii) Enhancing and strengthening information and knowledge concerning climate change vulnerabilities, and (iii) Raising awareness and coordination to enhance the adaptation to climate risks in development activities.

The Local Disaster Risk Reduction Facility (LDRRF) is another component of the CDMP. It works at the local level and helps in coordinating between development and management of disaster-related projects and programmes of the government. These mechanisms help in management of disasters and risks occurring due to climate change.

It has been demonstrated that the government integrates climate change issues into policies and programmes at the national level. The government has incorporated the issues of climate change into the management programmes of the coastal zone on the recommendation of the World Bank. It has also adopted programmes for disaster preparedness and sectoral plans for water.

The Poverty Reduction Strategy Paper[18] (PRSP) establishes a direct connection between natural disasters and poverty and also opines that the natural disasters would subsequently increase due to global warming. However, it does not specify the issue of climate change in planning different development measures (Kelkar & Badhwal, 2007). The Bangladesh government in 2007 announced that it would incorporate climate change issues into all projects of the PRSP and also proposed a plan of action and policy drought by the month of October in 2008 (Kelkar & Badhwal, 2007).

The National Water Policy (NWP) was announced by the government in 2001 to address various possibilities for water resources at all levels. Subsequently, the National Water Management Plan (NWMP) was announced in 2001, focussing on investment and implementation of issues prioritised in the NWP. Climate change is mentioned in the NWP, but the NWMP specifically recognises climate change and rising sea levels as one of the factors affecting the supply of water. The NWP and NWMP recommend flood-proof systems and early warnings of such risks. There are other policies related to the environment, like National Environmental Management Plan (NEMAP), the National Forest Policy, and National Land Use Policy, but these policies do not directly refer directly to the issues of climate change.

Bangladesh signed the United Nations Framework Convention on Climate Change (UNFCCC) in 1994. In Bangladesh, the Ministry of Environment and Forest (MOEF) coordinates the process and issues concerning the UNFCCC. A National Climate Change Committee (NCCC) was constituted in 1994 to look after the implementation and workings of the UNFCCC. It consists of members from government and non-government organisations (UNFCCC, 2022).

The Inter-Ministerial Committee on Climate Change (IMCCC) has representatives from Ministries, Non-Governmental Organisations, and research institutes. The Ministry of Environment and Forest looks into the matter of IMCCC. The National Environment Committee (NEC) determines the policies related to the environment. The Prime Minister of Bangladesh chairs the committee, which is represented by ministers and members from NGOs and civil society. Several studies have been conducted, individually or collectively, by various academic institutions, research organisations, universities, and government departments on vulnerabilities, impacts, and adaptation to climate change.

The Department of Environment in 2006 discussed national efforts to deal with issues of climate change in 'Addressing Climate Change in Bangladesh: National Efforts'. Bangladesh has participated in the US Climate Change Country Study Programme and prepared its assessment concerning climate issues in 1994. It also participated in the Asia Least Cost Greenhouse Abatement Strategy (ALGAS) study in 1958–98. The first National Communication to UNFCCC was submitted by Bangladesh in 2002. Subsequently, it has also taken the project, 'Bangladesh: Climate Change Enabling Activity' to prepare for the Second National Communication (UNFCCC, 2022).

Bangladesh has also submitted its National Adaptation Plan of Action (NAPA) in 2005 to the UNFCCC. A Designated National Authority (DNA) has been established at two levels by Bangladesh under the scheme of Clean Development Mechanism (CDM). These are the National CDM Board and National CDM Committee. There are four projects that have been approved by the DNA in the energy and waste sectors in Bangladesh. For example, in Matuail, it has focused on gas extraction by waste concern, and in Gazipur and Kanchpur, it has focused on composting projects. Through the project of Gramin Shakti, about three thousand solar systems have been installed, and energy efficient lamps have been encouraged in rural parts of Bangladesh. Bangladesh is the first Least Developed Country (LDC) that completed its National Adaptation Programmes of Action (NAPAs). This is related to papers and documents produced by LDCs to UNFCCC to specify their urgent requirements for adapting to climate change. The document emphasises six sectors like (i) Biodiversity, forestry, and land use, (ii) Livestock, fisheries, and agriculture, (iii) Health, natural disasters, coastal areas, and water (iv) Local governance, food security, gender, and livelihood (v) Infrastructure and industry, (vi) Policy and institutional issues.

The UNFCCC was signed during the United Nations Conference on Environment and Development, also known as the Earth Summit. It was held in 1992 in Rio de Janeiro. It addresses issues concerning global warming. It provides funds under four categories to deal with climate issues (UNFCCC, 2022).

1. Least Developed Countries Fund: It helps developing countries in preparing and implementing their National Adaptation Programmes of Action (NAPA, 2022).
2. Special Climate Change Fund: It supports different climate change activities, and provides technology and adaptation to climate change facilities.
3. Green Climate Fund: It is an operating entity of the financial mechanism of the convention.
4. Global Environment Facility: This provides financial assistance to different projects and activities concerning climate issues.
5. Adaptation Fund: This was established to provide financial assistance to programmes and projects of developing countries that face difficulties due to issues of climate change.

Though it has been found by several scholars that the UNFCCC fundings have been inadequate to meet the requirements of developing countries (Moehner & Klein, 2007), Bangladesh completed its NAPA project by receiving financial assistance from the Least Developed Countries Fund (LDCF) of the UNFCCC. However, there is a requirement for more financial assistance to carry out the full process of documentation of projects and their implementation.

Apart from this, Bangladesh receives development assistance of over one billion USD annually. An analysis of such donations was done by the Organisation for Economic Cooperation and Development (OECD) in the year 2005 (Agrawala et al., 2003). It concluded that about 22 to 53 per cent of development assistance and 22 to 37 per cent of funded projects are in climate risk-affected sectors (Agrawala et al., 2003). In 1996, the World Bank talked about the economic consequences of the rise in sea level in Bangladesh. The World Bank also sponsored the Bangladesh Climate Change and Sustainable Development Study in 2000. It analysed the consequences of climate change, took note of institutions that adapt to climate issues, and reviewed different projects for development like the National Water Management Plan. The aim is to analyse the development strategies adopted in Bangladesh. The OECD has concluded that the documents and strategies of the donor countries lack focus on the issues of climate change (Agarwala et al., 2003). However, with an increase in the vulnerabilities of climate change, there has been increase in focus on issues of climate change vulnerabilities in Bangladesh.

The Department for International Development (DFID, 2023) in Bangladesh has given priority to climate change issues in its programmes for development. It also supports projects and programmes that would reduce the effects of vulnerabilities of climate change. It is also working on projects related to climate change vulnerabilities, issues of disaster risk reduction, and adaptation to climate change. The adaptation measures have been adopted by the Department of Development Programmes, such as building good quality roads and infrastructure during floods in Bangladesh, and the education and awareness programmes for adaptation to climate change and disaster risk management. Different research studies and data collection efforts on the issues of impacts of climate change have been conducted. It also provides funds for climate change and disaster management. A total fund of 5.7 million pounds for five years has been provided by the DFID (Lockwood et al., 2006). The United Nations Development Programme (UNDP) and the Coastal and Wetland Biodiversity Management at Hakaluki Haor and Cox's Bazar (2000–2007) are other programmes that emphasise the effects of the rise in sea level (Agarwala, 2003).

Recommendations and Conclusion

The efforts at multiple levels would be needed to deal with challenges which would require finance, technology, and capacity building. People feel that natural disasters are divine tests as they have firm faith in religion. As such, (i) Awareness and knowledge regarding the issue can play an important role in dealing with climate change vulnerabilities. (ii) The findings that are scientific and indigenous related to climate change should be integrated and applied in the decision-making process. (iii) The timely and proper forecasting of weather would reduce the impacts of bad weather conditions. (iv) The seasonal migration[19] can play an important role in reducing and minimising the bad effects of weather conditions. (v) The conservation of water resources may be helpful during droughts and floods in Bangladesh. (vi) There is a need for proper adoption and implementation of policies and programmes that would minimise and reduce the risks of natural disasters. (vii) The conservation of common resources would be helpful during the scarcity of water, food, and medicine caused by natural disasters. (viii) Inter cropping[20] and Crop Rotation[21] can play significant roles in dealing with food insecurity due to climate change vulnerability. (ix) The focus should also be on soil conservation[22] and water management. (x) Social networks and platforms like Facebook, YouTube, WhatsApp, etc., can play an important role in raising awareness and circulating information on risks caused by climate change issues among the people. (xi) Cattle grazing pastures need to be managed to overcome the issues of climate change. (xii) There should be proper institutions to manage the issues of refugees, migration, and displaced persons. (xiii) The behavioural pattern of the people concerning resource management needs to be monitored. (xiv) The capacity-building capability of the government, institutional framework, and good governance at the local and national levels need to be ensured for benefiting the country. (xv) A proper channel of communication is required at different levels with different institutions for effective implementation of programmes and policies. (xvi) The capacity building and awareness programmes, amongst people together with its adaptation at all levels, be it local or national, are required.(xvii) The feedback of the people and other institutions is also important so that the government can measure the effectiveness of steps taken to overcome crises caused by climate change. (xviii) Migration leads to the integration of several ethnic, racial, and religious groups. This can increase tensions and violence among different groups. The government should play an important role and introduce programmes that help foster mutual understanding, tolerance, and resolution of conflicts in particular areas. The government, together with Non-Governmental Organisations and local governments, should initiate and implement such programmes for conflict management and resolution. (xix) Extremist ideology should be challenged and pluralism respected. The government should develop programmes and policies that restrict extremist ideologies and popularise non-violence. The education system should be developed in such a way that cultural harmony and pluralism are respected. (xx) People who migrate from coastal and rural areas to urban slums due to climate change vulnerabilities have agricultural skills and find themselves unemployed in urban areas. As such they find themselves on the verge of criminal practices, child labour, and trafficking. The people marry their young daughters at a young age to reduce economic burden. The government should play an important role and develop placement programmes concerning jobs that can be helpful to these people and reduce illegal economic sources. (xxi) Capacity-building programmes and the rule of law should be strengthened so that migrant populations are easily integrated by the government. (xxii) As migration also goes beyond borders, regional cooperation should be developed among countries of a particular region. The agreements, policies, and programmes should be developed so that the impacts of climate change are reduced and the problems of displaced people are addressed. (xxiii) Alternative means of livelihood should be ensured for small-scale businesses and people involved in agriculture and aquaculture. (xxiv) Proper healthcare facilities and access to clean and safe drinking water should be guaranteed for vulnerable populations. (xxv) Specific importance to vulnerable populations like women and children needs to be given and their capacities to adapt to climate change crises should be strengthened.

Notes

1 The three major rivers are Ganga, Brahmaputra, and Meghna.
2 The Sundarbans is a mangrove area in the delta formed by the confluence of the Ganges, Brahmaputra, and Meghna Rivers in the Bay of Bengal.
3 A flash flood is the rapid flooding of low-level areas, which are caused mostly by heavy rains.
4 Coastal flooding is the submerging of dry and low-lying land by seawater.
5 The process of settling down heavier insoluble particles at the bottom of a liquid is called sedimentation.

6 Salinisation is an increase in the total dissolved solids (TDS) of the aquifer caused by natural or anthropogenic factors.
7 The pathogens are microorganisms or bacteria that can cause diseases.
8 The increased temperature on the surface of the sea and ocean in the Eastern and Central Tropical Pacific Ocean.
9 The pressure group is a special interest group that seeks to influence the policy of the government in a particular direction.
10 A financial constraint is something that prevents the course of economic issues and actions.
11 The lack of enough resources to meet the demands and outcomes of a project.
12 The low-income countries confronting several structural impediments to sustainable development. They are highly vulnerable to economic and environmental shocks and have low levels of human assets.
13 The systemic corruption and lack of openness and accountability, arbitrary policymaking, and deception of those who are ruled.
14 The individual or group that has an interest in any decision or activity of an organisation.
15 The capacity of an organisation to be effective, productive, and relevant.
16 These are non-profit entities independent of governmental influence.
17 BIPSS is a non-party, non-profit organisation and an independent think tank that studies peace and security issues concerning the regions of South Asia and Southeast Asia.
18 The Poverty Reduction Strategy Papers are documents required by the International Monetary Fund and World Bank before countries are considered relieved from debt within the Heavily Indebted Poor Countries (HIPC) initiative.
19 The movement of people from one place to another on a seasonal basis.
20 The practice of growing two or more crops in proximity is called Intercropping.
21 The practice of growing different crops on a single soil to prevent soil exhaustion.
22 The prevention of erosion and depletion of soil from damage caused by water and wind.

References

Agrawala, S., Ota, T., Ahmed, A. U., Smith, J., & van Aalst, M. (2003). *Development and climate change in Bangladesh: Focus on coastal flooding and the Sunderbans*. Organisation for Economic Co-operation and Development (OECD).

Ahmad, Q. K., & Ahmed, A. U. (2002). *Bangladesh: Citizens' perspectives on sustainable development*. Bangladesh Unnayan Parishad (BUP).

Ahmad, Q. K., Ahmed, A. U., Karim, Z., Prasad, K., & Poudel, S. N. (2004). Community based flood management in South Asia. *Asia Pacific Journal on Environment & Development, 12*, 1–2.

Ahmad, Q. K., Ahmed, A. U., Khan, H. R. & Rasheed, K. B. S. (2001). 'GBM Regional Water Vision: Bangladesh Perspectives', in Q. K. Ahmad, A. K. Biswas, R. Rangachari & M. M. Sainju (Eds.), Ganges-Brahmaputra Meghna Region: A Framework for Sustainable Development, University Press Limited, Dhaka, pp. 31–80.

Ahmed, A. U. (2005) Adaptation options for managing water related extreme events under climate change regime: Bangladesh perspectives. In M. M. Q. Mirza & Q. K. Ahmad (Eds.), *Climate change and water resources in South Asia*. Balkema Press, pp. 255–278.

Ahmed, A. U. (2006). *Bangladesh: Climate change impacts and vulnerability, a synthesis*. Comprehensive Disaster Management Programme.

Ahmed, A. U., & Haque, N. (2002, March 14–18). *Climate change and sustainable development* [Paper presentation]. Dhaka Meet on Sustainable Development, Dhaka.

Ahmed, A. U., & Mirza, M. M. Q. (2000). Review of causes and dimensions of floods with particular reference to flood '98: National perspectives. In Q. K. Ahmad, A. K. A. Chowdhury, S. H. Imam, & M. Sarker (Eds.), *Perspectives on flood 1998*. The University Press Limited.

Asaduzzman, M., et al. (2005). *Climate change and Bangladesh: Livelihoods issues for adaptation*. Bangladesh Institute for Development Studies (BIDS)

BBS. (2020). *Statistical pocketbook of Bangladesh: 2019*. Bangladesh Bureau of Statistics (BBS), Government of the People's Republic of Bangladesh (GOB).

BIPSS. (2023). *BIPSS launches climate change cell*. Bangladesh Institute of Peace and Security Studies. https://bipss.org.bd/unit-climate-change-cell/

Department of Economics. (2006). *Climate change and implications for Bangladesh-input to discussion on Swedish cooperation strategy with Bangladesh* [Policy brief]. https://sidaenvironmenthelpdesk.se/digitalAssets/1683/1683274_climate-change-bangladesh-policy-brief-060315.pdf

DFID. (2022). Department for International Development, Bangladesh. https://www.commonwealthofnations.org/organisations/dfid_bangladesh___department_for_international_development/

Eckstein, D., et al. (2021). *Global Climate Risk Index 2021, who suffers most from extreme weather events? Weather related loss events in 2019 and 2000–2019*. GERMANWATCH.

Glennon, R. (2017). The unfolding tragedy of climate change in Bangladesh. *Scientific American*. https://blogs.scientificamerican.com/guest-blog/the-unfolding-tragedy-of-climate-change-in-bangladesh/

Global Climate Risk Index 2021. (2021). *Who suffers most from extreme weather events? Weather-related loss events in 2019 and 2000 to 2019*. Germanwatch. https://www.germanwatch.org/en/19777

GOB. (2005). *National Adaptation Programme of Action (NAPA), final report: November 2005*. Ministry of Environment and Forest, Government of the People's Republic of Bangladesh (GOB).

GED. (2018). Sustainable Development Goals: Bangladesh Progress Report 2018. Bangladesh Planning Commission, General Economics Division (GED). Dhaka: Government of the People's Republic of Bangladesh.

Haider, R. (1992). *Cyclone '91 revisited*. Bangladesh Centre for Advanced Studies.

Huq, S. (2001). *Climate change and Bangladesh*. International Institute for Environment and Development. https://www.researchgate.net/publication/11634613_Climate_Change_and_Bangladesh

IDMC. (2022). https://www.internal-displacement.org/countries/bangladesh

IPCCC. (2001). *Intergovernmental Panel on Climate Change, annex B, glossary of terms*. http://www.ipcc.ch/pdf/special-reports/srex/SREX-Annex_Glossary.pdf

Isfat, M., & Raihan, A. (2022). Current practices, challenges, and future directions of climate change adaptation in Bangladesh. *International Journal of Research Publication and Reviews, 3*(5), 3429–3437.

Karim, Z., Hussain, S. G., & Ahmed, A. U. (1998). Climate change vulnerability of crop agriculture. In S. Huq, Z. Karim, M. Asaduzzaman, & F. Mahtab (Eds.), *Vulnerability and adaptation to climate change for Bangladesh*. Kluwer Academic Publishers, pp. 39–54.

Karim, Z., Hussain, S. G., & Ahmed, M. (1996). Assessing impacts of climate variations on foodgrains production in Bangladesh. *Journal of Water, Air and Soil Pollution, 92*, 53–62.

Kelkar, U., & Bhadwal, S. (2007). *South Asian regional study on climate change impacts and adaptation: implications for human development: Human* development report *2007–2008,* https://cdn.cseindia.org/challenge_balance2012/pdf/Adap_South%20Asia%20HDR%202007%20READ.pdf
Khatun, F., & Saadat, S. Y. (2021). *Climate change in Bangladesh: A sustainable development perspective.* Centre for Policy Dialogue.
Koelle, K., Pascual, M., & Yunus, M., (2005). Pathogen adaptation to seasonal forcing and climate change. *Biological Sciences,* 272(1566), 971–977.
Lockwood, H., Schoen, P., & Taher, M. (2006). *DFID-Bangladesh: Output to purpose review of comprehensive disaster management programme.* United Nations Development Fund.
Mahtab, F. (1989). *Effect of climate change and sea level rise on Bangladesh, (for expert group on climate change and sea level rise).* Commonwealth Secretariat.
McCarthy, J. J., Canziani, O. F., Leary, N. A., Dokken, D. J., & White, K. S. (eds). (2011). Work group II input to the third assessment report. In J. J. McCarthy et al. (Eds.), *Climate change 2001: Impacts, adaptation and vulnerability: contribution of Working Group II to the third assessment report of the Intergovernmental Panel on Climate Change (Vol. 2).* Cambridge University Press.
Ministry of Environment and Forest. (2009). *Government of the People's Republic of Bangladesh, National Adaptation Programme of Action.* https://unfccc.int/resource/docs/napa/ban02.pdf
Mirza, M. M. Q., Dixit, A., & Nishat, A. (2003). *Flood problem and management in South Asia.* Kluwer Academic Publishers.
Moehner, A., & Klein, R. (2007). *The global environment facility: Funding for adaptation of adapting to funds?* [Climate and Energy Working Paper]. Stockholm Environment Institute (SEI).
NAPA. (2022). *National Adaptation Programmes of Action, United Nations Climate Change.* https://unfccc.int/topics/resilience/workstreams/national-adaptation-programmes-of-action/introduction
Nizamuddin, K., Khuda, Z. R. M. M., & Ahmad, N. (2001). Women-headed households displaced by river bank erosion: Problems and strategies of survival. In K. Nizamuddin (Ed.), *Disaster in Bangladesh: Selected readings.* Disaster Research Training and Management Centre.
Rahman, A., & Alam, M. (2003). *Mainstream adaptation to climate change in Least Developed Countries (LDCS)* (Working Paper No. 2). International Institute for Environment and Development.
Ramamasy, S., & Bass, S. (2007). *Climate variability and change: Adaptation to drought in Bangladesh, a resource book and training guide.* Food and Agriculture Organization of the United Nations
Rodó, X., Pascual, M., Fuchs, G., & Faruque, A. S. G. (2002). ENSO and cholera: A nonstationary link related to climate change? *Proceedings of the National Academy of Sciences, 99,* 12901–12906.
RVCC. (2003). *Report of a community level vulnerability assessment conducted in Southwest Bangladesh.* Reducing Vulnerability to Climate Change (RVCC) Project, CARE Bangladesh.
Siddiqi, M. H. (1997, October 8). *Floods: Problems and prospects* [Paper presentation]. Workshop on observance of the International Decade for Natural Disaster Reduction 1997, Organized by the Disaster Management Bureau, Ministry of Disaster Management and Relief, The Government of the People's Republic of Bangladesh (GoB), Dhaka, Bangladesh.
Siddique, Q. I., & Chowdhury, M. M. H. (2000). Flood '98: Losses and damages. In Q. K. Ahmad, et al. (Eds.), *Perspectives on flood 1998.* The University Press Limited (UPL).

UNO (2022). UNDESA. https://www.un.org/development/desa/dpad/least-developed-country-category/ldc-graduation.html
WEF (2023). Climate change is costing the world $16 million per hour: study, World Economic Forum. https://www.weforum.org/stories/2023/10/climate-loss-and-damage-cost-16-million-per-hour/#:~:text=Using%20this%20method%2C%20the%20team,severity%20over%20the%20long%20term
The World Bank. (2022). *Data bank, world development indicators.* https://databank.worldbank.org/reports.aspx?source=2&country=BGD
The World Bank. (2022). *Data: Bangladesh.* https://data.worldbank.org/country/bangladesh?view=chart
UNFCC. (2022). *United Nations climate change, introduction to climate action.* https://unfccc.int/climate-action/introduction-climate-action
United Nations. (2022). *LDC portal-international support measures for least developed countries: Bangladesh graduation status.* https://www.un.org/ldcportal/content/bangladesh-graduation-status
WB. (2000). *Bangladesh: Climate change and sustainable development* (Report No. 21104-BD). Rural Development Unit, South Asia Region, The World Bank.

Index

Note: Page references in **bold** reference tables; page references with "n" followed by numbers represents note numbers.

FSC
www.fsc.org
MIX
Papier | Fördert gute Waldnutzung
FSC® C083411

Kolibri 360 GmbH
Ferdinand-Jühlke-Straße 7
99095 Erfurt, Deutschland
produktsicherheit@kolibri360.de